INTELLIGENT COMPONENTS FOR AUTONOMOUS AND SEMI-AUTONOMOUS VEHICLES

(ICASAV'95)

*A Postprint volume from the IFAC Workshop,
Toulouse, France, 25 - 26 October 1995*

Edited by

P. BIDAN
LAAS du CNRS, Toulouse, France

and

S. BOVERIE
Siemens Automotive S.A., Toulouse, France

Published for the

INTERNATIONAL FEDERATION OF AUTOMATIC CONTROL

by

PERGAMON
An Imprint of Elsevier Science

UK   Elsevier Science Ltd, The Boulevard, Langford Lane, Kidlington, Oxford, OX5 1GB, UK

USA   Elsevier Science Inc., 660 White Plains Road, Tarrytown, New York 10591-5153, USA

JAPAN   Elsevier Science Japan, Tsunashima Building Annex, 3-20-12 Yushima, Bunkyo-ku, Tokyo 113, Japan

First edition 1996

Library of Congress Cataloging in Publication Data

A catalogue record for this book is available from the Library of Congress

British Library Cataloguing in Publication Data

A catalogue record for this book is available from the British Library

ISBN 0-08-042603 4

Printed and bound in Great Britain by
CPI Antony Rowe, Chippenham and Eastbourne

IFAC WORKSHOP ON INTELLIGENT COMPONENTS FOR AUTONOMOUS AND SEMI-AUTONOMOUS VEHICLES

Sponsored by
International Federation of Automatic Control (IFAC)
- Technical Committee on Components and Instruments

Organized by
LAAS-CNRS
Siemens Automotive S.A.
SITEF - CCIT

Co-sponsored by
IFAC Technical Committees on
- Automotive Control
- Intelligent Autonomous Vehicles
- Aerospace--
French National Member Organization (AFCET)

International Programme Committee
A. Ollero (E)   (Chairman)
F. Assbeck (D)
P. Bidan (F)
S. Boverie (F)
A. Casals (E)
A.G. Cerezo (E)
P. Dario (I)
D. Esteve (F)
A. Halme (SF)
C.J. Harris (UK)
R. Isermann (D)
M. Jamshidi (USA)
R.A. Jarvis (AA)

U. Kiencke (D)
G. Montseny (F)
D. Pomerleau (USA)
H. Roth (D)
G. Sandini (I)
A. Stenz (USA)
C. Thorpe (USA)
A. Titli (F)
M. Tomizuka (USA)
G. Ulivi (I)
M. Valentin (F)
W.H. Wimmer (D)

National Organizing Committee
P. Bidan   (Chairman)
S. Boverie
A. Dreuil
M.T. Ippolito
F. Vernières

CONTENTS

AUTOMOTIVE
Driver Assistance

AUTONOMOUS VEHICLES AND ENERGY MANAGEMENT

Integrated Vehicle Control Systems

U. Kiencke
University of Karlsruhe
Institute for Industrial Information Systems
Hertzstr. 16, D-76187 Karlsruhe
Fax: 0049 - 721 - 755788
email: kienke@iiit.uni-karlsruhe.de

Keywords:

ABS-Braking, Networking, Yaw-control, Misfiring Detection, Optimal Routing

Abstract:

Control Systems have penetrated Automobiles in the last twenty years, improving exhaust emissions, driving safety and comfort.
By local networks, such dedicated functions may now be integrated into an overall vehicle control system, requiring advanced modelling, estimation and control procedures. In this paper, an overview about integrated vehicle control systems is presented.

1. Introduction

Starting about 25 years ago, dedicated functions in automobiles have been enhanced by electronic components and by control loops. Engine management systems were derived from simple ignition modules, in which the dwell time was closed-loop controlled and the mechanical ignition switch was substituted by a power transistor. With the event of the emission legislation and the three way catalyst, air-fuel ratio control was developed. The air and the evaporated fuel takes a delay time to go through the combustion chamber and the exhaust pipe. This delay depends on engine speed and engine load. The regulator performs a limit cycle, with a period proportional to the delay time, and with an amplitude proportional to the ratio of delay time to integration time of the controller. In order to overcome long air-fuel mismatches during engine transients, a learning adaptation was introduced [1].

By assigning mismatches to additive and multiplicative errors, a global offset compensation could be found, which significantly improves emission levels, especially during transients.

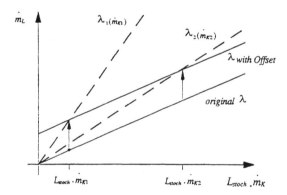

Figure 1: Global Lambda Offset Compensation

Another important dedicated control system was ABS braking. In emergency braking, drivers usually press down the braking pedal as far as they can, thereby generating the maximum hydraulic pressure within the brakes. The friction coefficient between the wheels and the road is first increasing with mounting slip, until a maximum is passed. Beyond that point, friction starts to decrease, destabilizing the braking system. Wheels rapidly decelerate until they block entirely. With rather simple means, ABS braking systems detect this developing instability by monitoring wheel deceleration. In case of light decelerations, braking pressure is lowered, returning the wheels to the stable region before the friction maximum [2].

1

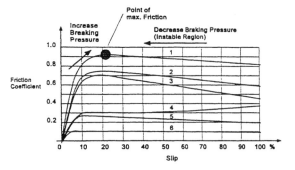

Figure 2: ABS - Breaking

In the resulting slip oscillations, the ABS control system automatically finds the approximate region of maximum friction, thereby shortening braking distances.

Such dedicated control functions operated isolated from each other. Rather simple plant models were sufficient as a basis for controller design.
The main reason was the lack of an economic communication link for automotive applications.

2. Networking

In 1986, Controller Area Network (CAN) was first presented [3]. In this local area network, control modules are regarded as autonomous agents, which cooperate without any master or other centralized coordination. Errors are detected and corrected by renewed transmission. Failures are collectively classified and assigned to a local station, which might be cut off to allow for a degraded system function. Distributed safety-relevant control systems can thus be supported.

Even more important for control applications are the following features.

- Priorisation of messages
  Time constants in control systems vary in a wide range. When messages in control loops with small time constants get high message priorities, the network transfer capacity is well adapted to control requirements.

- Contention based arbitration
  Distributed control systems are event-driven by nature. E.g. the exceeding of a speed threshold may be an event, which activates a speed limitation procedure in a remote module. Another example is the synchronization of concurrent processes. Messages are generated by such events, which then contend for access to the network.

- Locality

  If a locality measure L is defined as the ratio of

  $$L = \frac{\text{Message Transmission Time}}{\text{Network Propagation Time}},$$

  the local coherence of networks can be quantified. For a high locality

  $$L \gg 1,$$

  network propagation can be neglected against message transmission, i.e. the location of the message source becomes insignificant ($L_{CAN}=10^3$). In control applications, there is no restriction from the network on feedback or intensive information exchange. Control functions may therefore be arbitrarily placed in a vehicle.

The resulting logical integration of all former dedicated control functions by a network with high locality offers a new platform for the application of modern control science.
Control functions are no longer arbitrarily restricted to certain electronic control units. Rather, the entire drivetrain or vehicle dynamics can now be modelled and controlled with a uniform approach. Thus automatic control evolves as a key innovation driver in Automotive Systems.

3. Integrated Vehicle Control

A first example shall be yaw control of vehicles. In order to have a vehicle cornering, side slip angles of the tyres are required for building up lateral tyre forces. The vehicle body turns to a yaw angle, by which side slip angles are distributed among front and rear tyres.
A non-ideal distribution leads to the well known phenomena of over- and understeering.
For correction of the yaw angle, wheels at one side of the vehicle can be braked for a short time [4]. Another command input is rear wheel steering, as shown in [5].

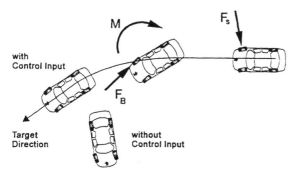

Figure 3: Yaw Control in case of oversteering

A sliding mode control establishes a fixed dynamic relationship between the drivers steering command and the vehicles yaw rate. That approach eliminates the impact from the wide range of car and road parameters and from varying operating conditions, preserving the steady state behavior of conventional cars.

Vehicle Dynamics are thus not just taken for granted, but actively improved by automatic control. Kiencke, Dais [6] presented a real-time estimation of the tyre friction. By modelling the braking and vehicle dynamics, and by measuring braking pressures, wheel speeds and vehicle acceleration, the friction characteristic can be derived. This allows to adapt braking to the actual conditions of the road surface. By estimating the entire friction characteristic versus wheel slip rather than just the actual friction coefficient, braking performance can be extrapolated beyond the momentary operating point. While the estimation process takes about half a second for settling, a subsequent extrapolation does not require any delays for constant road surfaces. Other examples for the estimation of vehicle state variables is the determination of the yaw angle of the vehicle body. By combination of such modelling, estimation procedures and advanced control algorithms shall significantly improve the driving performance of vehicles.

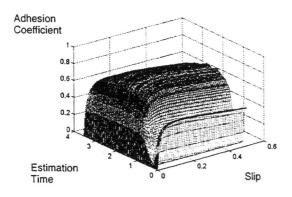

Figure 4: Estimation of Adhesion Characteristic

Another important area is diagnostics. Mayer [7] presented a comparative approach to detect pressure drops in single tyres that is based exclusively on the measurement of the four wheel speeds. At a given vehicle speed, the individual wheel speeds are determined by the tyre effective rolling radius, which changes in dependence of tyre pressure. The ratio of rotational speeds can be estimated for all four wheels. In case of straight vehicle movements, this ratio deviates by about 0.2% at a pressure drop of 0.5 bar. The problem in this approach is, that cornering of the vehicle increases the speed of the outer wheels, and that vehicle acceleration increases the speed of the driven wheels.

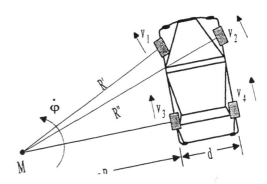

Figure 5: Vehicle Cornering

These disturbing influences are larger than that of the tyre pressure. One approach could be to estimate disturbances and to correct wheel speeds. Fuzzy estimation has proved to be an efficient tool for that. In comparison to classical Kalman filtering, signal uncertainties can be much better determined by heuristic experience. The procedure shows, that even in very special application areas, the behavior of the entire vehicle must be known, when advanced algorithms shall be applied.

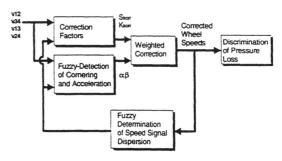

Figure 6: Tyre pressure diagnostics by Fuzzy Estimation

Another well known diagnostic problem is the misfire detection in spark ignition engines for the onboard diagnosis of the emission control system.

Henn [8] has presented a Kalman filter to estimate in-cylinder pressure torque from the engine speed signal. The nonlinear time dependent model is transformed into a linear angular based model which serves as a base for Kalman filtering.

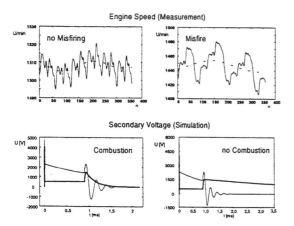

Figure 7: Misfire detection

Krishnaswami, Rizzoni [9] proposed a nonlinear autoregressive moving average model in order to detect various engine faults. For misfire detection, the spectrum of the angular velocity signal in the order of rotation domain is calculated. When misfire occurs, the energy in the frequency components comprised between the order of engine cycle and the firing frequency increases. The misfire index then contains the phase information of the misfiring cylinder, which can be discriminated by a so-called geometric classifier.

Automobiles are thus regarded as an entity, and their modelling now contains all the physical behaviour required for a respective control or diagnostic scheme.

4. Outlook

Beyond the limits of a single vehicle, the dynamic relationship of many cars and their information exchange can be the target of automatic control. As part of the American "Intelligent Vehicle Highway System" (IHVS) program or the European Prometheus program, the longitudinal and lateral behavior of vehicles shall be controlled. By putting several vehicles together into a so-called platoon, the given capacity of existing highways can be better used. In long platoons with low inter-vehicle distances, stability can only be achieved by intensive information exchange between the vehicles [10].

Another step could be the optiomal routing of individual vehicles from their starting point to the destination through a highway network. Conventional routing problems may be handled e.g. by considering the road network as a graph, where the arrows are weighted by the expected driving times.

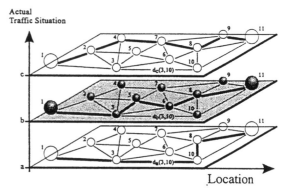

Figure 8: Optimal routing depending on traffic situation

In each node, the optimal residual path to the destination can be found. However, there are two problems in such an approach.

I. While proceeding on the once calculated optimal path from one node to it's successor, the overall traffic condition might have changed. Expected driving times may no longer be optimal under the actual situation. Had one known this change in advance, a different routing would have been adopted. One approach to handle such problems is to introduce the so-called identical probability for situation changes, and weight all other path options with it.
II. The access to certain routes can be curtailed in case of exceeding demand. By selecting a non optimal route for part of the drivers, the entire traffic network can achieve optimal performance.

Such systems are modelled as discrete event systems, which are currently evolving as a new branch in control theory.

5. Summary

Control systems in vehicles used to concentrate on dedicated applications. With the introduction of real-time networks, vehicles are now approached as integrated control systems. Their complexity and nonlinear behavior are a challenge to control engineers worldwide. Beyond control of single vehicles, multi-vehicle control and traffic management systems might be a next step.

6. Literature

[1]   Kiencke, U. "A View of Automotive Control Systems", IEEE Control Systems Magazine, Vol. 8 No. 4, Aug. 1988, pp 11–19

[2]   Leiber, H. ; Czinzel, A. "Antiskid System for Passenger Cars with Digital Electronic Control Unit", 1979, SAE-Paper 790458

[3]   Kiencke, U. ; Dais, S. and Litschel, M. " Automotive Serial Controller Area Network", SAE Congress Feb 24–28, 1986, SAE-Paper 860391

[4]   Zanten, A. "FDR - Die Fahrdynamikregelung von Bosch" ATZ Automobiltechnische Zeitschrift 96, 1994

[5]   Guldner, J. ; Utkin, V. I. ; Ackermann, J. and Bünte, T. "Sliding Mode Control for active steering of Cars" IFAC-Workshop Advances in Automotive Control, March 13–18, 95, Monte Verita, Ascona, Switzerland

[6]   Kiencke, U. and Daiß, A "Estimation of Tyre Friction for enhanced ABS–Systems", International Symposium on Advanced Vehicle Control AVEC, Japan, 1994, pp. 515–520

[7]   Mayer, H. "Comparative Diagnosis of Tyre Pressures", 3rd IEEE Conference on control applications, 24–26 Aug., 1994, Glasgow, pp. 627–632

[8]   Henn, Michael "Estimation of In-Cylinderpressure Torque from Angular Speed Kalmann Filtering", IFAC-Workshop Advances in Automotive Control, March 13–18, 95, Monte Verita, Ascona, Switzerland pp. 20–25

[9]   Krishnaswami, V. and Rizzoni, G. "Onboard diagnosis of engine faults", IFAC-Workshop Advances in Automotive Control, March 13–18, 95, Monte Verita, Ascona, Switzerland pp. 138–143

[10] Hedrick, J. K. "Vehicle Control Issues in Intelligent vehicle Highway Systems", IFAC-Workshop Advances in Automotive Control, March 13–18, 95, Monte Verita, Ascona, Switzerland, pp. 189–196

THE CRONE SUSPENSION

X. Moreau, A. Oustaloup and M. Nouillant

Equipe CRONE - LAP - ENSERB - Université Bordeaux I
351, cours de la Libération - 33405 Talence Cédex - FRANCE
Tel. (33) 56.84.61.40 - Fax. (33) 56 84 66 44
E-mail : oustalou@lap.u-bordeaux.fr

Abstract: This paper deals with robust control and advanced suspension of vehicles, and more specifically, a new system called CRONE suspension based on non integer derivation. The CRONE suspension results from a traditional suspension model whose spring and damper are replaced by an actuator defined by a non integer transmittance. This system is called CRONE suspension thanks to the link with the second generation CRONE control, i.e. the vertical template. That is why the principle of the second generation CRONE control is used to synthesise the CRONE suspension transmittance. The suspension parameters are determined from a constrained optimisation of a performance criterion. A two degree of freedom quarter car model is used to evaluate performances. The frequency and time responses, for various values of the vehicle load, reveal a great robustness of the degree of stability through the constancy of the resonance ratio in the frequency domain and of the damping ratio in the time domain.

Keywords: robust control, second generation CRONE control, advanced suspension, CRONE suspension, robustness of the degree of stability.

1. INTRODUCTION

Automotive suspensions are designed to provide good vibration insulation of the passengers and to maintain adequate adherence of the wheel for braking, accelerating and handling.

The introduction of electronic in automotive suspensions has been considered for decades but it is only recently that the automotive industry has begun to seriously consider modulated, semi-active and active suspensions. Many reports (Rakheaja, 1991), (Yasuda, 1991), (Jezequel, 1992) have shown that the introduction of active or semi-active elements in suspension increases vehicle performance, even given the ride comfort/road-holding ability dilemma (Hedrick, 1990), using an optimal approach (Daver, 1990), (Moreau, 1993). Improvement has been obtained not only thanks to progress in hydraulics and electronics (Oustaloup, 1990), (Dunwoody, 1991), (Gohring, 1993) but also thanks to control law research on active or semi-active elements of suspension (Redfield, 1989), (Yue, 1989).

This paper deals with a robust control law for a suspension system which develops a force which is proportional to the non integer derivative of its relative displacement. This system is called CRONE suspension thanks to the link with the second generation CRONE control (Commande Robuste d'Ordre Non Entier) (Oustaloup, 1991) i.e the template which characterizes both the control and the suspension.

The paper is divided into five parts. Part I gives the principle of the CRONE suspension and its model. Part II develops the synthesis method of the suspension. Part III describes the constrained optimisation to determine CRONE suspension parameters. Part IV examines the frequency and time responses which show the robustness of both the resonance and damping ratios versus load variation. Finally, in part V conclusions are given.

2. MODELING

The quarter car model shown in Fig. 1 has been studied by many authors to analyse and optimise automotive suspensions.

7

The CRONE suspension results from a traditional suspension model whose spring and damper are replaced by an actuator (Fig. 1) defined by a non integer order transmittance. m_2 is the mass supported by each wheel and taken as equal to a quarter of the total mass of the body. k_2 is the stiffness of the spring and b_2 the damping coefficient for a traditional suspension. $C(s)$ is the CRONE suspension transmittance replacing the traditional suspension transmittance. k_1 is the stiffness and b_1 the damping coefficient of the tyre. m_1 is the unsprung mass. $z_0(t)$ is the deflexion of the road, $z_1(t)$ and $z_2(t)$ are the vertical displacements of the wheel and body respectively.

The suspension develops a force $F_2(s)$ which is a function of the relative displacement $Z_{12}(s)$ and which obeys symbolically to the general relation :

$$F_2(s) = C(s)\, Z_{12}(s) , \qquad (1)$$

with

$$Z_{12}(s) = Z_1(s) - Z_2(s) . \qquad (2)$$

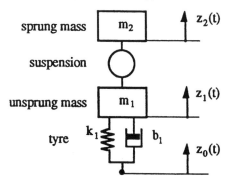

Fig. 1. Two degree-of-freedom model

If it is assumed that the tyre does not leave the ground and that $z_1(t)$ and $z_2(t)$ are measured from the static equilibrium position, then the application of the fundamental law of dynamics leads to the linearised equations of motion :

$$m_1\, \ddot{z}_1(t) = f_1(t) - f_2(t) \qquad (3)$$

and

$$m_2\, \ddot{z}_2(t) = f_2(t) , \qquad (4)$$

in which

$$f_1(t) = k_1 \left(z_0(t) - z_1(t) \right) + b_1 \left(\dot{z}_0(t) - \dot{z}_1(t) \right) \qquad (5)$$

and $f_2(t)$ the force developed by the suspension.

The Laplace transform of equations (3), (4) and (5), assuming zero initial conditions, are

$$m_1\, s^2\, Z_1(s) = k_1 \left(Z_0(s) - Z_1(s) \right) \\ + b_1\, s \left(Z_0(s) - Z_1(s) \right) - F_2(s) \qquad (6)$$

and

$$m_2\, s^2\, Z_2(s) = F_2(s) , \qquad (7)$$

or, given equation (1) of $F_2(s)$,

$$m_1\, s^2\, Z_1(s) = k_1 \left(Z_0(s) - Z_1(s) \right) \\ + b_1\, s \left(Z_0(s) - Z_1(s) \right) - C(s) \left(Z_1(s) - Z_2(s) \right) \qquad (8)$$

and

$$m_2\, s^2\, Z_2(s) = C(s) \left(Z_1(s) - Z_2(s) \right) . \qquad (9)$$

To analyse the vibration insulation of the sprung mass, two transmittances are defined :

$$T_2(s) = \frac{Z_2(s)}{Z_1(s)} = \frac{C(s)}{m_2\, s^2 + C(s)} \qquad (10)$$

and

$$S_2(s) = \frac{Z_{12}(s)}{Z_1(s)} = \frac{m_2\, s^2}{m_2\, s^2 + C(s)} . \qquad (11)$$

To study ride comfort and road holding ability, three additional transmittances are defined :

$$H_a(s) = \frac{A_2(s)}{V_0(s)} , \; H_{12}(s) = \frac{Z_{12}(s)}{V_0(s)} , \; H_{01}(s) = \frac{Z_{01}(s)}{V_0(s)}, \qquad (12)$$

in which $A_2(s)$ is acceleration of the sprung mass, $Z_{12}(s)$ suspension deflection, $Z_{01}(s)$ tyre deflection and $V_0(s)$ road input velocity. A commonly used road input model is that $v_0(t)$ is white noise whose intensity is proportional to the product of the vehicle's forward speed and a road roughness parameter.

3. SYNTHESIS METHOD OF THE CRONE SUSPENSION

The synthesis method of the CRONE suspension is based on the interpretation of transmittances $T_2(s)$ and $S_2(s)$ which can be written as :

$$T_2(s) = \frac{\beta(s)}{1 + \beta(s)} \qquad (13)$$

and

$$S_2(s) = \frac{1}{1 + \beta(s)}, \qquad (14)$$

in which

$$\beta(s) = \frac{C(s)}{m_2\, s^2} . \qquad (15)$$

The transmittances $T_2(s)$ and $S_2(s)$ can here be considered to be of an elementary control loop whose $\beta(s)$ is the open loop transmittance.

Given that relation (15) expresses that a variation of mass is accompanied by a variation of open loop gain, the principle of the second generation CRONE control (Oustaloup, 1991) can be used by synthesising the open loop Nichols locus which traces a vertical template for the nominal mass.

3.1 First version of the CRONE suspension

A first way of synthesising the Nichols locus, defined in Fig. 2, consists in determining an open loop transmittance which presents an asymptotic behaviour of order n' between 1 and 2.

This asymptotic behaviour can be obtained with a transmittance of the form :

$$\beta(s) = \left(\frac{\omega_u}{s} \right)^{n'} . \qquad (16)$$

Identification of equations (15) and (16) leads to :

$$\beta(s) = \left(\frac{\omega_u}{s} \right)^{n'} \qquad (17)$$

in which

$$n = 2 - n' \in \;]0,1[\qquad (18)$$

and

$$\omega_0 = \frac{1}{\left(m_2 \, \omega_u^{2-n}\right)^{1/n}} \; . \qquad (19)$$

Equation (1) becomes then :

$$F_2(s) = \left(\frac{s}{\omega_0}\right)^n \left[Z_1(s) - Z_2(s)\right] \; , \qquad (20)$$

namely, in the time domain :

$$f_2(t) = \frac{1}{\omega_0^n} \left(\frac{d}{dt}\right)^n \left[z_1(t) - z_2(t)\right] \; . \qquad (21)$$

Equation (21) thus obtained expresses that the CRONE suspension develops a force which is proportional to the non integer derivative of its relative displacement. The non integer order is between 0 and 1 (Oustaloup, 1993).

3.2 Second version of the CRONE suspension

Another way of synthesising the open loop Nichols locus, given that robustness does not require an infinitely long template, consists in determining a transfer β(s) which successively presents (Fig.3) :
- an order 2 asymptotic behaviour at low frequencies to eliminate tracking error ;
- an order n' asymptotic behaviour where n' is between 1 and 2, exclusively around frequency ω_u to limit the synthesis of the non integer derivation at a truncated frequency interval ;
- an order 2 asymptotic behaviour at high frequencies, to ensure satisfactory filtering of vibrations at high frequencies.

Such localised behaviour can be obtained with a transmittance of the form :

$$\beta(s) = C_0 \left(\frac{1 + \frac{s}{\omega_b}}{1 + \frac{s}{\omega_h}}\right)^n \left(\frac{\omega_0'}{s}\right)^2 \qquad (22)$$

in which :

$$\omega_b \ll \omega_A \; , \; \omega_B \ll \omega_h \; \text{and} \; n = 2 - n' \in \;]0,1[\; . \; (23)$$

Identification of equations (15) and (22) gives :

$$1/\sqrt{m_2} = \omega_0' \qquad (24)$$

and

$$C(s) = C_0 \left(\frac{1 + \frac{s}{\omega_b}}{1 + \frac{s}{\omega_h}}\right)^n \; . \qquad (25)$$

The equation obtained defines the ideal version of the suspension. The corresponding real version (Oustaloup, 1991) is defined by a transfer of integer order N :

$$C_N(s) = C_0 \prod_{i=1}^{N} \left(\frac{1 + \frac{s}{\omega_i'}}{1 + \frac{s}{\omega_i}}\right) \; , \qquad (26)$$

in which :

$$\frac{\omega_{i+1}'}{\omega_i'} = \frac{\omega_{i+1}}{\omega_i} = \alpha\eta > 1 \; ; \; \frac{\omega_i'}{\omega_i} = \alpha \; ; \; \frac{\omega_{i+1}}{\omega_i'} = \eta \; ;$$

$$\alpha\eta = \left(\frac{\omega_h}{\omega_b}\right)^{1/N}; \; \alpha = (\alpha\eta)^n \; ; \; \omega_1' = \omega_b \, \eta^{1/2} \qquad (27)$$

and $\omega_N = \omega_h \, \eta^{-1/2}$.

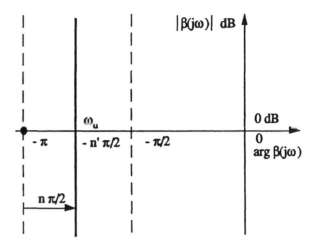

Fig. 2. Open loop Nichols locus of the first version CRONE suspension

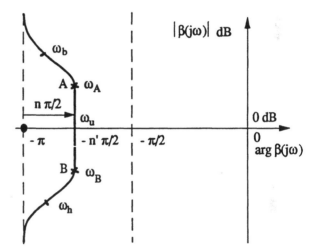

Fig. 3. Open loop Nichols locus of the second version CRONE suspension

4. DETERMINATION OF CRONE SUSPENSION PARAMETERS : CONSTRAINED OPTIMISATION

By defining the transmittances (12) with respect to $v_0(t)$, all frequencies contribute equally to their mean square values. That is why the determination of CRONE suspension parameters is based on the minimisation of a criterion J composed of the H_2-norm of the transmittances $H_a(j\omega)$, $H_{12}(j\omega)$ and $H_{01}(j\omega)$, namely :

$$J = \frac{\rho_1}{\lambda_1} \int_{\omega_b}^{\omega_h} |H_a(j\omega)|^2 \, d\omega + \frac{\rho_2}{\lambda_2} \int_{\omega_b}^{\omega_h} |H_{12}(j\omega)|^2 \, d\omega$$

$$+ \frac{\rho_3}{\lambda_3} \int_{\omega_b}^{\omega_h} |H_{01}(j\omega)|^2 \, d\omega + \frac{\rho_4}{\lambda_4} \int_{\omega_b}^{\omega_h} |H(j\omega)|^2 \, d\omega, \qquad (28)$$

in which ρ_i are the weighting factors, λ_i the H_2-norm computed for the traditional suspension and $H(j\omega)$ the transmittance between force $F_2(j\omega)$ developed by the suspension and the road input velocity $V_0(j\omega)$, namely :

$$H(j\omega) = \frac{F_2(j\omega)}{V_0(j\omega)} = m_2 \, H_*(j\omega) \; . \quad (29)$$

To obtain a significant comparison between traditional and CRONE suspension performances, two constraints are fixed for the minimal sprung mass :
- equal unit gain frequency of open loop $\beta(j\omega)$;
- equal resonance ratio of transmittance $T_2(j\omega)$.

For the traditional suspension, the expression of unit gain frequency ω_{u2} and resonance ratio Q_2 are given by

$$\omega_{u2} = \frac{\sqrt{b_2^2 + \sqrt{b_2^4 + 4 \, m_2^2 \, k_2^2}}}{\sqrt{2} \, m_2} \quad (30)$$

and

$$Q_2 = \frac{2 \sqrt{2} \, \zeta_2^2}{\sqrt{\sqrt{1 + 8 \, \zeta_2^2} - 1 - 4 \, \zeta_2^2 + 8 \, \zeta_2^4}} \; , \quad (31)$$

in which $\quad \zeta_2 = \dfrac{b_2}{2\sqrt{k_2 \, m_2}} \; . \quad (32)$

For the CRONE suspension, the expression of unit gain frequency ω_u and resonance ratio Q are given by (Moreau, 1995)

$$\omega_u = \frac{1}{(m_2 \, \omega_0^n)^{1/(2-n)}} \quad (33)$$

and

$$Q = \frac{1}{\sin\left[(2-n) \dfrac{\pi}{2}\right]} \; . \quad (34)$$

5. PERFORMANCES

5.1 Example

The traditional suspension is a rear suspension of a 605 Peugeot whose parameters are :

- sprung mass : $\qquad 150 \text{ kg} \le m_2 \le 300 \text{ kg}$;
- stiffness : $\qquad k_2 = 18\,000 \text{ N/m}$;
- damping coefficient : $\qquad b_2 = 2450 \text{ Ns/m}$;
- unsprung mass : $\qquad m_1 = 24 \text{ kg}$;
- stiffness of tyre : $\qquad k_1 = 273\,820 \text{ N/m}$;
- damping coefficient of tyre : $\qquad b_1 = 50 \text{ Ns/m}$;

From this data, the constrained optimisation of the criterion J, computed with the optimisation toolbox of Matlab, provides the optimal parameters of the first version CRONE suspension, namely :

$$n = 0.763 \quad \text{et} \quad \omega_0 = 2.032 \; 10^{-5} \text{ rd/s} \; . \quad (35)$$

The optimal parameters of the second version CRONE suspension are :
- for the ideal version :

$$n = 0.763 \; ; \qquad C_0 = 2\,667 \; ;$$
$$\omega_b = 0.628 \text{ rd/s} \; ; \; \omega_h = 314 \text{ rd/s} \; ; \quad (36)$$

- for the real version :

N = 5 ; $\qquad C_0 = 2\,667$;

$\alpha = \omega_i/\omega_i' = 3.272$; $\qquad \eta = \omega_{i+1}'/\omega_i' = 1.445$;

$\omega_1' = 0.7553 \text{ rd/s}$; $\qquad \omega_1 = 2.471 \text{ rd/s}$;

$\omega_2' = 3.572 \text{ rd/s}$; $\qquad \omega_2 = 11.69 \text{ rd/s}$; $\qquad (37)$

$\omega_3' = 16.89 \text{ rd/s}$; $\qquad \omega_3 = 55.3 \text{ rd/s}$;

$\omega_4' = 79.87 \text{ rd/s}$; $\qquad \omega_4 = 261.3 \text{ rd/s}$;

$\omega_5' = 314.7 \text{ rd/s}$; $\qquad \omega_5 = 1235.8 \text{ rd/s}$.

5.2 Frequency responses

Figs. 4, 5 and 6 show frequency performances in open loop and in closed loop.

Fig. 4 gives the Nichols loci $\beta(j\omega)$ for the traditional and CRONE suspensions. The phase margin varies with mass m_2 for the traditional suspension. On the other hand, phase margin is independent for the CRONE suspension, where the Nichols loci in open loop trace the template which characterizes the second generation CRONE control.

Fig. 5 and 6 gives the gain diagrams of $T_2(j\omega)$ and $S_2(j\omega)$ for traditional and CRONE suspensions. For the CRONE suspension, the resonance ratio can be seen to be both weak and insensitive to variations of mass m_2. This shows a better robustness of the CRONE suspension in frequency domain.

5.3 Step responses

Fig. 7 and 8 shows the step responses of the car body and the wheel for both suspensions. For the CRONE suspension it can be seen that the first overshoot remains constant, showing a better robustness for the CRONE suspension in the time domain (fig. 7.b). The road holding ability is the same for both suspensions (fig. 8).

6. CONCLUSION

In this paper it has been shown that the CRONE suspension provides remarkable performances : better robustness of stability degree versus load variations of the vehicle. Robustness is illustrated by the frequency and time responses obtained for different values of the load.

It is shown that this robustness is due to the template which characterizes implicitly the CRONE suspension. This template characterizes the second generation CRONE control explicitly.

From the concept of the CRONE suspension two technological solutions have been developed (Moreau, 1995). The first, called *passive CRONE suspension*, uses the link between recursivity and non integer derivation. This suspension has now been mounted on an experimental Citroën BX. The second solution, called *passive piloted CRONE suspension*, uses a continuously controlled damper. Its design permits manufacture at the traditional automobile damper cost. Bench tests on a prototype have validated theoretical expectations.

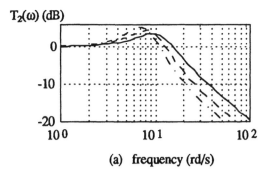

(a)   frequency (rd/s)

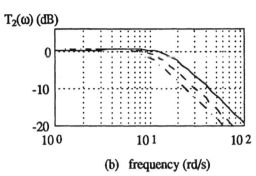

(b)   frequency (rd/s)

Fig. 5. Gain diagrams of $T_2(j\omega)$ for traditional (a) and CRONE (b) suspensions :
(————) m_2 = 150 kg ; (– – – – –) m_2 = 225 kg ;
(– · – · –) m_2 = 300 kg

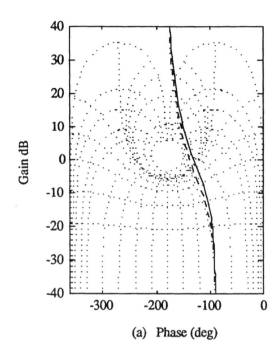

(a)   Phase (deg)

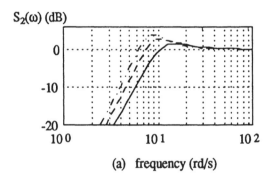

(a)   frequency (rd/s)

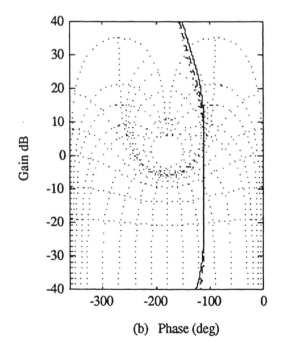

(b)   Phase (deg)

Fig. 4. Nichols loci in open loop for (a) traditional and (b) CRONE suspensions :
(————) m_2 = 150 kg ; (– – – – –) m_2 = 225 kg ;
(– · – · –) m_2 = 300 kg

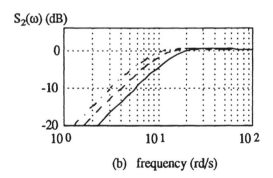

(b)   frequency (rd/s)

Fig. 6. Gain diagrams of $S_2(j\omega)$ for traditional (a) and CRONE (b) suspensions :
(————) m_2 = 150 kg ; (– – – – –) m_2 = 225 kg ;
(– · – · –) m_2 = 300 kg

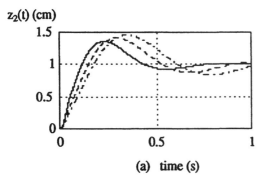

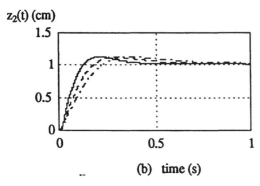

Fig. 7. Step responses of sprung mass for traditional (a) and CRONE (b) suspensions :

(————) m_2 = 150 kg ; (— — — —) m_2 = 225 kg ;

(— · — · —) m_2 = 300 kg

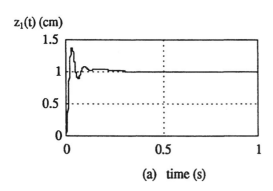

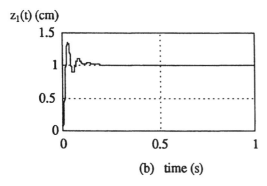

Fig. 8. Step responses of wheel for traditional (a) and CRONE (b) suspensions :

(————) m_2 = 150 kg ; (— — — —) m_2 = 225 kg ;

(— · — · —) m_2 = 300 kg

REFERENCES

Daver R., Roberti V. and Jezequel L. (1992). Nouvelle stratégie de suspension semi-active. *SIA*, n°672, pp.57-62.

Dunwoody A. B. (1991). Digital active suspension. *SAE Transactions*, vol.100, n°6, pp.1649-1659.

Gohring E. and Von Glasner (1993). Intelligent suspension for commercial vehicles. *Proceedings of the International Congress MV2*, Active Control in Mechanical Engineering, tome 1, pp.1-12.

Hedrick J.K. and Butsuen T. (1990). Invariant properties of automotive suspensions. *Journal of automobile engineering*, Vol.204, n°1, pp.21-27.

Jezequel L. and Roberti V. (1992). Behaviour of a preview semi-active suspension. In: *Mécanique-Matériaux-Electricité*, n°445, pp.32-35.

Moreau X., Oustaloup A. and Nouillant M. (1993). La suspension CRONE : une suspension active d'ordre non entier optimal. *Proceedings of the International Congress MV2*, Active Control in Mechanical Engineering, tome 1, pp.79-92.

Moreau X. (1995). Intérêt de la dérivation non entière en isolation vibratoire et son application dans le domaine de l'automobile. La suspension CRONE : du concept à la réalisation. Thèse de Doctorat, Université Bordeaux I.

Oustaloup A. and Nouillant M. (1990). Nouveau Système de Suspension, Brevet n°9004613, INPI.

Oustaloup A. (1991). La Commande CRONE, HERMES ED.

Oustaloup A., Moreau X. and Nouillant M. (1993). From the second generation CRONE control to the CRONE suspension. *Proceedings of the International Congress IEEE*, International Conference on Systems, Man and Cybernetics, Le Touquet, tome 2, pp.143-148.

Redfield R.C. and Karnopp D.C. (1989). Performance sensitivity of an actively damped vehicle suspension to feedback variation. *Journal of dynamic Systems, Measurement, and Control*, Vol.111, n°1, pp.51-60.

Rakheaja S. and Ahmed A. K. W. (1991). Simulation of non-linear variable dampers using energy similarity. *Journal of Engineering Computation*, vol.8, n°4, pp.333-344.

Yasuda E. and Doi S. (1991). Improvement of vehicle motion and riding-comfort by active controlled suspension system. *SAE Transactions*, vol. 100, n°6, pp.890-897.

Yue C., Butsuen T. and Hedrick J.K. (1989). Alternative Control Laws for Automotive Active Suspensions. *Journal of dynamic Systems, Measurement, and Control*, Vol.111, n°2, pp.286-291.

A FUZZY-LOGIC BASED INTELLIGENT SUSPENSION WITH CONTINUOUSLY VARIABLE DAMPING

C.F. Nicolás†, J. Landaluze†, X. Sabalza†, M. Gastón‡ and R. Reyero†

†IKERLAN, Control Engineering Department, 20500 Arrasate-Mondragón, Spain
‡LIPMESA-QUINTON HAZELL, 01400 Llodio, Spain

Abstract: This paper presents an intelligent suspension system consisting of a continuously variable shock absorber and an Electronic Control Unit which implements control strategies developed on the basis of fuzzy logic. These take into account the actions of the driver. Due to the simplicity of the sensor system required, it can be used in any type of vehicle. The system, called LIPTRONIC, has been implemented in a number of vehicles. The results achieved with the fuzzy controller are at a similar level to those normally achieved with other much more expensive systems based on vehicle dynamics.

Keywords: Intelligent suspension, electronic suspension, continuously variable shock absorber, semi-active suspension, fuzzy control.

1. INTRODUCTION

A passive suspension system is a compromise between comfort and safety. Depending on the choice of spring or shock absorber, one or other concept is stressed. How hard the shock absorber is determines the dynamics between the wheel and the load and, therefore, driving safety.

In order to try and overcome this compromise, in the 70s and early 80s the concepts of active and semi-active suspension were developed (Karnopp, 1983), although due to technological limitations they did not get beyond the academic or theoretical level. Throughout the 80s variable suspension systems, called semi-active suspension, electronic suspension or intelligent suspension systems, started to be used in commercial vehicles. It was hoped that there would be an explosion in the use of these systems in the 90s, similar to that which occurred at the end of the last decade with ABS systems. Although the expectations have been delayed due to the world economic recession, the appearance of these systems in mid-range commercial vehicles, like the Ford

Mondeo or the Citroën Xantia, has reinforced the popularity of these concepts.

In the last few years, Ikerlan has been working on semi-active suspensions (Landaluze et al., 1990), and has developed and tuned up, together with the shock absorber manufacturer Lipmesa-Quinton Hazell, several prototypes, based on both the actions of the driver and vehicle dynamics, or on a combination of the two. A basic element of the suspension system carried out is the continuously variable shock absorber which Lipmesa-Quinton Hazell developed several years ago and has recently perfected for the LIPTRONIC project. This enables the design of high performance suspension systems (Katsuda et al., 1992).

This paper describes an intelligent suspension system, developed preferably on the basis of the actions of the driver. Fuzzy logic techniques have been applied to take advantage of the continuous characteristics of the Lipmesa-Quinton Hazell shock absorber. The system is simple, although from the point of view of results, the performance is similar to that obtained from other much more costly and complex systems based on vehicle dynamics.

Likewise, other systems using three damping rates have been prepared, with algorithms based on threshold values and tables with regard to the speed of the vehicle from the variables observed, which have served as a reference. The systems developed have been implemented in a Renault R-11 vehicle specially prepared for testing suspension systems, as well as a Land Rover Discovery vehicle, with which more than 20,000 miles of tests have been carried out. The results obtained are presented in this paper.

2. SYSTEM DESCRIPTION

The intelligent suspension system presented here consists of the following basic elements: the shock absorbers, the sensors and the Electronic Control Unit. The main lines of the comfort/safety strategy were established on the basis of these elements.

2.1 Shock absorbers

The main element is the variable shock absorber, which, to a great extent, determines the specifications of a semi-active suspension system.

In the Lipmesa-Quinton Hazell variable shock absorber, the oil flow between the compression chamber and the reverse chamber of the shock absorber is controlled by an electrovalve. This enables the damping forces to be varied continuously in both expansion and compression, depending on the current applied to the electrovalve. Its main features from the point of view of its structure are as follows:

- Externally it is the same as a standard one. To avoid changes in its structure, the oil passes through a counter-rod.
- It is designed in twin-tube technology.
- It consists of standard parts, with a proportional electrovalve located at the top of the rod, in the low frequency part of the shock absorber.
- The output of the electrovalve supply wires is through the inside of the piston rod.

The force-speed laws vary continuously depending on the intensity passing through the electrovalve, between 0 and 0.8 A. The response time is 10 ms. A non-linear model was built to characterize the continuously variable shock absorber and serve as the basis for the implementation of certain control algorithms. For example, Figure 1 shows the mathematical model for a rear shock absorber in the Renault R-11 vehicle, on the basis of the experimental tests carried out by the manufacturer.

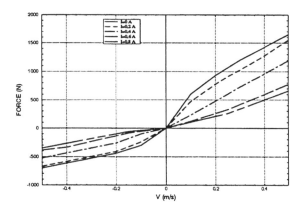

Fig. 1. Shock absorber simulation model

2.2 Sensors

If the vehicle supplies a digital signal proportional to the speed, which is what happens in most vehicles nowadays, the intelligent suspension system only needs two sensors to be specifically installed: one to determine the steering wheel angle and another to determine the throttle valve position (or alternatively the position of the accelerator pedal). Moreover, the system uses the brake pedal switch, as well as an accelerometer used in the Electronic Control Unit circuit to measure the vertical accelerations of the sprung mass.

2.3 Electronic Control Unit

A preindustrial Electronic Control Unit has been developed for general and modular purposes. It enables all types of algorithms to be implemented, from the simplest ones based on the actions of the driver to the most complicated ones based on vehicle dynamics. It has been built using SMD technology and the latest recommendations at EMC level for car equipment have been taken into account.

The electronic card includes the hardware necessary for conditioning and offsetting the temperature of an accelerometer, so that this can be included in the card as an additional component.

The most important specifications of the ECU can be summarized as follows: based on the SAB80C166 microcontroller, it has 32K of RAM, 32K of EPROM and 16x16 bits of EEPROM. Inputs: 10 analog channels and 10 digital channels. Outputs: 4 independent current outputs for the shock absorbers, RS-232-C serial communications port, possible CAN bus and digital channels for information LEDs.

With this hardware, in the algorithms described in this paper, the variables sensed and the control loop are run every 10 ms.

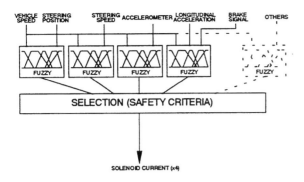

SELECTION (SAFETY CRITERIA)

SOLENOID CURRENT (x4)

Fig. 2. Fuzzy selection criteria

2.4 Safety/Comfort strategy

Information from the sensors continuously reaches the Control Unit, which decides which is the corresponding operating point of the variable shock absorber, always as part of the compromise between safety and comfort. By default, priority is given to comfort, with the shock absorbers operating in the softest areas, although from the variables considered a switch is made to harder values, giving more weight to safety over comfort, when it is required.

The different values sensed or estimated affect some of the four components of the control strategy: ride, handling (including roll control), acceleration and deceleration (dive). The variables considered and their effects are as follows:
- The speed of the vehicle. A basic variable to determine the driving conditions and to interpret the values of the other variables.
- The steering wheel angle position. This is used to detect when the vehicle is cornering. The damping force must be determined from the centrifugal acceleration value estimated depending on the vehicle speed and the steering wheel angle.
- The speed of the steering wheel. This enables emergency manoeuvres to avoid obstacles or collisions to be detected, in cases in which it is necessary both to increase the adhesion of the tyres and limit vehicle roll.
- The position of the throttle valve. The power which the driver requires of the engine is determined, enabling preventive actions to be taken to avoid squat effects.
- Brake pedal pressure. Enables dive phenomena to be avoided.
- Vertical acceleration of the sprung mass. To a certain extent, this enables information about vehicle dynamics to be obtained and the type of road or existing problems to be detected. It also enables oscillation phenomena of the sprung mass to be avoided, which are harmful from both the comfort and vehicle safety points of view.

3. CONTROLLER DESIGN

Three types of controllers have been developed and implemented with the variables observed and estimated: one based on threshold values, another based on tables and a fuzzy controller. In the first two only three discrete levels of damping force are considered in the shock absorber. In the fuzzy controller, the shock absorber works as continuously variable and can take any value between the extreme limits of the damping level.

In the three types of controller, the driver can choose between automatic and manual mode by means of a switch. Manual mode determines a fixed rating value for the shock absorbers, in the hard position for most vehicles, as is fitting for sports driving. However, this strategy and the operating logic behind the shock absorber electrovalve changes for off-road vehicles, for which a soft damping value is established, suitable for driving on unsurfaced roads.

3.1 Control policy based on threshold levels

The control strategies based on threshold values for the variables observed are the most classic of those implemented in controllers which take into account the actions of the driver. The shock absorber basically works according to an ON/OFF strategy, switching between two levels, hard and soft. The intermediate value is only taken into account for very low speeds (v < 5 Km/h) or as another possibility in the changes caused by the vertical acceleration of the sprung mass.

3.2 Control policy based on tables

In the control strategy based on tables for each analog value observed or estimated, a table has been devised according to vehicle speed, with the output being one of the three discrete levels of damping considered. In principle, this enables the safety/comfort strategy to be adapted to different types of driving, which are determined by the speed of the vehicle anyway.

3.3 Fuzzy controller

The reason for developing a fuzzy control is based on the need to calculate a variable damping level in a continuous range according to the actions taken by the vehicle driver, in order to obtain optimum operation of the semi-active damping system. The fuzzy system is based on the definition of a set of rules and functions which associate the level of damping with specific driving conditions according to observations carried out by the sensors.

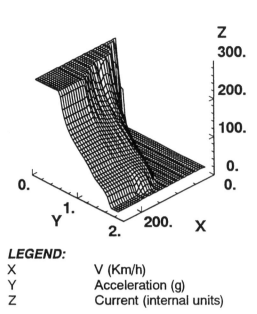

LEGEND:
X   V (Km/h)
Y   Acceleration (g)
Z   Current (internal units)

Fig. 3. Characteristic surface for the steering angle

Fig. 4. "Stewart" platform

Safety Criteria. Between multiple choices depending upon observation of different driving conditions, the one that guarantees the higher level of safety must be selected. Comfort is the control objective only when safety is not severely diminished. Due to the problems involved in incorporating negative reasoning (required when, due to the driving circumstances, the system has to maintain operational safety) in a purely fuzzy system, the controller obtained is actually a hybrid.

When designing a rule-based inference system, theoretically a general formulation depends on rules with as many antecedents as considered inputs. In practice this leads to a cumbersome system that needs a powerful CPU to compute the results in a restricted real-time environment. A more simple solution was conceived, based on a modular approach that allows many input variables and criteria for suspension operation to be incorporated (see Figure 2).

The actual value is selected by means of a selection algorithm that performs the following functions:
• When the suspension is about to be set at a softer level then some temporisation may be introduced via a low-pass discrete filter. This prevents undesirable softening when a succession of abrupt manoeuvres is performed.
• Selection of the hardest value for damping.

Fuzzy controller computational structure. The controller basically consists of three parts:
• Input pre-processing. Includes all the filtering and sensor signal processing required by the system.

• Fuzzy inference. Computations of fuzzy reasoning.
• Postprocessing (selection algorithm). Determines the damping value to be applied at each moment, bearing in mind the different values inferred.

Fuzzy controller rule base. In order to simplify the design of the control system a controller based on four different knowledge bases was formulated. Each rule base considers two antecedents and one consequent. One of the antecedents is always the longitudinal speed of the vehicle, and the other may be one of the following: steering wheel angle, steering wheel rotational speed, vertical acceleration of the vehicle and longitudinal acceleration (measured as accelerator pedal position or throttle valve position). In all cases the consequent is the damping level, expressed in attack current to the shock absorber electrovalve (the same for the front and rear shock absorbers).

The formulation of the system gives priority to safety over comfort in the driving conditions. By default, the state of the suspension will be that which guarantees a high level of comfort, although when driving circumstances require harder damping, to ensure the safety and stability of the vehicle during manoeuvres. For that, the damping selection criteria is: vehicle velocity vs. steering angle; vehicle velocity vs. steering speed; vehicle velocity vs. longitudinal acceleration; vehicle velocity vs. longitudinal deceleration; vertical acceleration.

Different linguistic variables are defined for the inputs, depending on the working ranges of the variables, in order to obtain a proper characterization of the driving conditions.

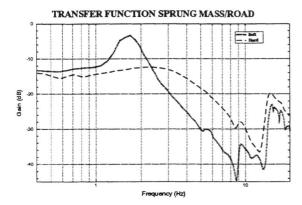

Fig. 5. Bode diagram of suspension system

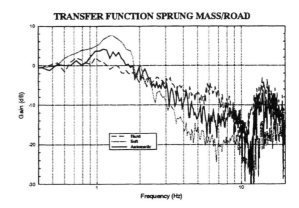

Fig. 6. Experimental frequency response of system based on threshold values

For the outputs only three linguistic terms are considered: SPORT, NORMAL and COMFORT, which refer to the hardest, an intermediate and the softest value. But, selecting Centre of Area method for defuzzyfication makes the working range of the output values continuous.

For each of the variables processed, and mentioned previously, a characteristic surface has been generated (rated in relation to the speed of the vehicle and the variable processed). For example, Figure 3 shows the characteristic surface created for the steering wheel position, in which an auxiliary antecedent is really being considered: the centripetal acceleration. This depends on the steering wheel position, the longitudinal speed and other characteristic parameters of the vehicle (distance between axles, demultiplication of the steering wheel).

Debugging. As the controller reacts to different actions taken by the driver, the effect of which on vehicle dynamics is not easily foreseen, it is virtually impossible to simulate the reaction of the system to all possible driving situations. Therefore a decision was made to partially validate the different submodules of the controller, carrying out the overall operational tests on the road.

Implementation of the fuzzy controller. The underlying idea is that in the final implementation of the system there should not be any variations in the knowledge base of the fuzzy system (adaptive characteristics have not been implemented).

Given that the fuzzy controller is a deterministic system, which faced with the same stimuli will always provide the same response, it is possible to precalculate the results of the fuzzy inference for a discrete set of values for the antecedents. This information is stored in table form (LUT), defined for a meshing of the antecedents. When certain values for the system inputs are observed in real time, this table is consulted, and it is determined between which values the output of the fuzzy system

will be. To calculate the approximate output which the fuzzy system will give us, a calculation by multilinear interpolation is carried out.

This procedure was implemented by means of integer arithmetic, ensuring a high calculation speed, although a prior analysis of the precision requirements is required when doing the calculations (overflows, divisions, etc).

4. TUNING UP THE SUSPENSION SYSTEMS

The tuning up of the system has been carried out at two levels: first of all the system was characterized and tuned up in the laboratory with a test platform. Then the final tests were carried out and the parameters were finely tuned on the road.

The activities carried out in the laboratory were based on a general purpose "Stewart" platform. This platform has 6 degrees of freedom and can withstand a weight of up to 2000 Kg, with a frequency response in a range of up to 15 Hz. It has been specially programmed and equipped with sensors so that semi-active suspension systems can be characterized, tuned up and tested, both those based on vehicle dynamics and those based on driver dynamics. Figure 4 shows a view of the test platform with a vehicle. The types of simple movements programmed are, for examples: frequency sweeps, simulations of real spectrums of different types of roads, simulation of avoiding obstacles, braking (anti-dive), problems on the road (holes and bumps), etc. These can all be combined together.

5. RESULTS

By way of an example of the type of tests carried out in the laboratory with the "Stewart" platform, Figure 5 shows the Bode diagram for the transfer function between the acceleration of the sprung mass and the input of the base or road. This figure shows the

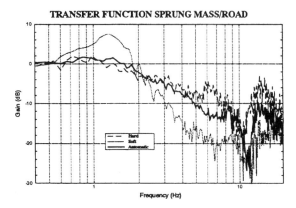

TRANSFER FUNCTION SPRUNG MASS/ROAD

Fig. 7. Experimental frequency response of system based on tables

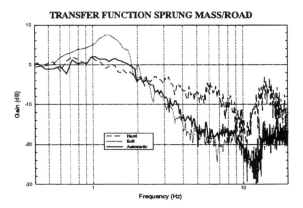

TRANSFER FUNCTION SPRUNG MASS/ROAD

Fig. 8. Experimental frequency response of system based on fuzzy logic

characterization of the hardest and softest positions of the continually variable shock absorber. This characterization serves as a reference for the dynamic performance which can be achieved with the best intelligent controller. The ideal situation is that, at low frequencies of up to about 2 Hz, it behaves like a hard suspension, to avoid big movements due to the resonance of the sprung mass and thus increases safety. From that frequency it should behave like a soft suspension.

As examples of the tests carried out on the road with the three control strategies implemented, Figures 6, 7 and 8 show the results obtained experimentally for the transfer function of the acceleration of the sprung mass with respect to the road input acceleration. These results correspond to driving at cruising speed on a dual carriage way, at an average speed of 100 Km/h. The acceleration of the sprung mass was measured in the copilot position and road input was approximated by measuring the acceleration in the axle of the corresponding wheel.

Figure 6 shows the results obtained with the shock absorber operating with three discrete values of damping coefficient and with the control strategy based on threshold values. Figure 7 shows the results obtained with a control strategy based on tables. These show a trend towards giving greater priority to safety over comfort. In any case, the main improvements in this system compared to the one based on threshold values is clear, above all, in turnings and swerves, as well as in better adaptation to different types of road.

Figure 8 shows the characteristics achieved with the fuzzy controller, with the shock absorber functioning as continuously variable. It can be seen that up to 2 Hz it practically behaves like a hard suspension system, and that from that frequency like a soft suspension system, with big improvements in the level of safety and comfort compared to conventional suspension systems. These results have been corroborated by other types of tests carried out, and

are similar to those achieved with semi-active suspension systems based on vehicle dynamics, which, in addition, require much more expensive sensors.

6. CONCLUSIONS

The connection of a continuously variable shock absorber to an Electronic Control Unit and the implementation of an algorithm based on fuzzy logic have enabled an intelligent suspension system with a level of performance unattained with previously developed systems. The improvement in the levels of safety and comfort have been confirmed by tests carried out in real conditions.

The fuzzy controller has been compared with others based on three discrete settings of the shock absorber damping coefficient, and demonstrates superior characteristics, in spite of using the same minimum level of sensors. For test purposes, these systems have been installed in a specially prepared Renault R-11 and a Land Rover Discovery. The resulting variable suspension system, LIPTRONIC, with a good cost/profit ratio can be adapted to all types of vehicles. The testing and tuning up of the parameters can be carried out on line by means of specially developed software tools.

REFERENCES

Karnopp, D. (1983). Active Damping in Road Vehicle Suspension Systems. *Vehicle System Dynamics*, **12**.

Katsuda, T., Hiraiwa, N., Doi, S. and Yasuda, E. (1992). Improvement of Ride Comfort by Continuously Controlled Damper. *SAE, No. 920276*.

Landaluze, J., Calzada, M. and Reyero, R. (1990). Aspectos de Control para Suspensiones Activas y Semiactivas. *Automática e Instrumentación*, No. **203**.

AUTOMOTIVE APPLICATIONS AND SIMULATIONS
OF A NEW HIGH BANDWIDTH SERVO-VALVE.

F.Guillemard*, J.Dore*, G.Dauphin-Tanguy**

*: *PSA Etudes et recherches,*
centre technique Citroën, route de Gisy, 78140 VELIZY VILLACOUBLAY.
**: *LAIL, Ecole Centrale de Lille,*
cité scientifique, BP48, 59651 VILLENEUVE D'ASCQ CEDEX

Abstract: A new multipurpose servo valve has been designed. Its direct-drive architecture enables high bandwidth, low cost and its use in automotive conditions. The validity of such affirmations has been verified on a test rig reproducing three main automotive applications. Simultaneously, a Bond-Graph model of the servo valve has been created. To verify its precision, it was inserted in the suspension test rig Bond-Graph model. Simulation and experience provide similar results. The potentiality of such actuator is demonstrated by the insertion of its model in a ¼ hydropneumatic active suspension model.

Keywords: servo valve, Bond-Graph, actuator, simulation, modeling, hydraulic

1. INTRODUCTION.

Numerous improvements have been done in electronics and data processing this last ten years. These improvements are about to be applied to vehicles. Future vehicles will integrate more and more intelligent functions able to counterbalance all sorts of outdoors and indoors car disturbances, assuring not only an optimal security, but also an optimal ride comfort and road handling.

But, as fast the speed of the controllers systems may be, so sophisticated the control algorithms could be, the final dynamic of these intelligent functions will be always imposed by the actuators dynamics. Having a real fast actuator is necessary. But this actuator must be able to transfer a large power too. And every engineer knows that hydraulic is the only form of energy which authorizes big transfer of energy in relatively small time and with small quantity of fluid (Guillon, 1991a). Consequently, the most suitable approach to develop an efficient function is to create and use an ultra fast electrohydraulic actuator or, in other words, a servo valve.

However, until now, numerous technical limitations to servo valve added to an high cost production prevent personal vehicles from having efficient control units able to assure a real security, ride comfort or road holding.

That is why, PSA and Thomson developed a new servo valve. Based on a direct-drive architecture (This evolution was forecast by Faisandier (1986) and Guillon (1991b)), it was designed to become a real interface between the control units and the transfer of big amount of energy and also in the aim of answering to the constraints of many different applications for the pressures and volumes used for these (Menard, 1994). Such robustness was verified using this servo valve for three different applications: The pressure control of a caliper brake, the position control of a steering wheel, and the pressure control of a hydropneumatic suspension. Numerous experimental results were obtained thanks to the creation of a test rig reproducing all these three applications. Robustness and celerity have been verified using this test rig.

Simultaneously, a study of the servo valve constitutive elements using Bond-Graph technique was made to create a simulation model. Bond-Graphs offer many advantages in terms of simplicity and efficiency. Their virtue lies in the relatively few symbols used to model dynamic systems and the possibility to associate many different energy domains. Bond-Graphs were the easiest way to model the servo valve, made with electronic, electrical, magnetic, mechanical and hydraulic components.

Once the identification made between the model and the reality, the servo valve model was inserted in the Bond-Graph model of the test rig suspension application in order to compare experimental results with simulation. The results obtained were correct and confirm the servo valve model accuracy.

Another model, more realistic, of a 1/4 hydraulic active suspension system was created including the model of the new servo valve. The comparison between responses of the system, including or not the actuator dynamic, to a disturbance force simulating turning maneuver or braking action was made. It demonstrates that the servo valve bandwidth is sufficient to obtain a correct suspension handling.

2. SERVO VALVE DESCRIPTION.

The servo valve aim is to create a link between an electrical signal coming from a controller and hydraulic power. As fig (1) shows, it is composed of two main elements :

    * A motor transforming the input control delivered by a PWM in mechanical position.
    * A spool valve transforming a linear position in hydraulic flow.

BRAKE CALIPER CONTROL

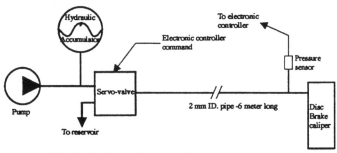

Fig. 2. Brake caliper test rig.

The axis of the motor is connected to the spool valve through a rod system, transforming a rotary into linear position of the spool.

To improve the efficiency of such system, an angular position sensor is used on the motor axis. This position information is used in the controller unit of the servo valve to control perfectly the angle position of the motor and consequently the spool position by comparing it with a control position input. Controlling the spool position, it is then possible to control numerous hydraulic components using this servo valve by controlling the load flow or the load pressure involved. For that, a second control unit must be used. This unit uses information of all sensors put on the system, decides on some action, and produces a position control input for the servo valve.

Such servo valve architecture is called direct drive. Its simplicity produces a high efficiency and robustness.

3. TEST RIG DESCRIPTION AND RESULTS.

The servo valve has been used for three applications different for the pressure and volume involved and for the state variables controlled. A test rig has been created demonstrating the advantages of such a servo valve for these three applications.

3.1. Pressure control of a caliper brake (fig (2)).

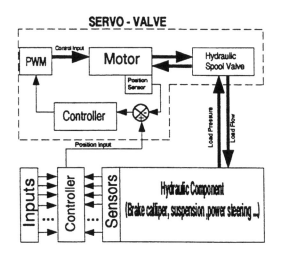

Fig. 1. Servo valve functionnal Description.

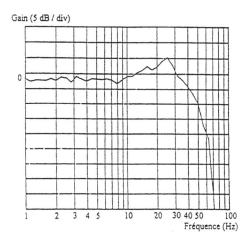

Fig. 3. Brake caliper Bode gain curve.

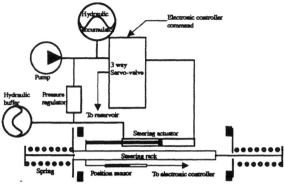

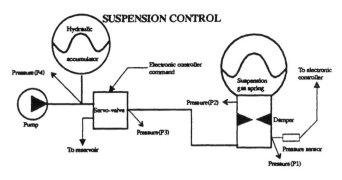

Fig. 4. Power steering test rig.

Fig. 6. Suspension test rig.

direction due to friction effects.

The load output of the servo valve is connected via a 6 meters long pipe to a disk brake caliper. An hydraulic accumulator gives a constant pressure to the servo valve. A pressure sensor is placed on the disk caliper. This pressure is compared to a pressure control input. The error between these two signals is used to calculate the command of the servo valve. The frequency response to sine command of 10 bars amplitude around 90 bars mean pressure is represented on fig (3). A bandwidth of 40 Hz is obtained for 3 dB attenuation.

3.2. Position control of a power steering (fig (4)).

The load output of the servo valve is connected to a steering actuator. The pressure regulator provides an initial pressure to the left chamber of the steering actuator. The pressure in the right chamber must give an equivalent effort to maintain the piston on stationary position. When the piston moves under the control of the servo valve, the hydraulic buffer prevents the system from any hydraulic jamming by admitting or providing additional flow to the left piston chamber. As a result, the pressure in that chamber changes in relation with the piston displacement. Additional springs on each side of the steering rack provide additional variation of effort function of the piston stroke. These efforts change in accordance with the steering actuator displacement

Thanks to a position sensor, the steering actuator can be controlled by the servo valve to follow a control position. Fig (5) shows the control input in comparison with the real actuator position. A triangular signal of 10.9 mm and 15 Hz frequency is applied. Despite a large variation of efforts applied on the steering rack function of the piston stroke, the real position is similar to the control position with only 25 ms delay, providing a 164 mm/s actuator speed or 1190°/s speed at steering wheel. This experience exhibits the high efficiency of the servo valve for the control of the piston motions despite large effort disturbances.

3. 3. Pressure control of suspension (fig (6)).

The servo valve load output is connected to a chamber associated with a hydraulic damper and hydropneumatic gas spring. An hydraulic accumulator provides constant pressure to the servo valve. To control the pressure P_1 in the chamber, a sensor is used to estimate this pressure.

An electronic controller compares the real P_1 pressure with the P_1 control pressure input and calculates the control servo valve position to apply for controlling P_1. Fig (7) shows the response of the suspension to a 30 bars step signal upwards and downwards.

Steering actuator control
Triangular signal

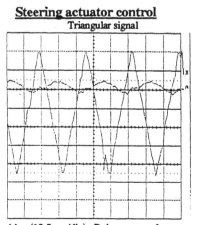

A: actuator position (12.5mm/div) , B: input control
Abscissa : time (0.025 second/div)

Fig. 5. Power steering test rig. Response to a triangular position signal.

Suspension Control
Pressure step signal

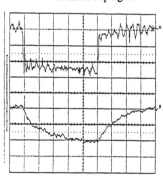

A: Pressure (P1) before damper (12.5 bars/div)
B: Pressure(P2) after damper(12.5 bars/div)
Abscissa : Time(0.2 second/div)

Fig. 7. Suspension test rig. Response to a pressure step signal.

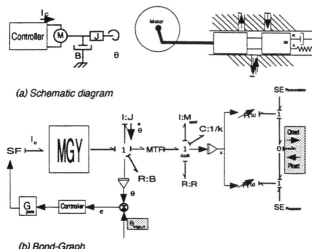

(a) Schematic diagram

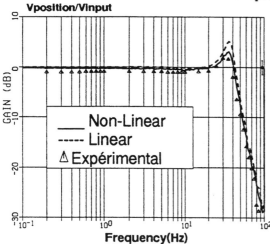

(b) Bond-Graph

Fig. 8. Servo valve Bond-Graph model

After about 4/100 s, the pressure P_1 and its control input are the same and remain the same despite the evolution of the pressure P_2 after damper. The spool valve position evolution shows the presence of a load flow smaller and smaller which regulates the pressure P_1 before damper during this period.

These three applications clearly demonstrate the robustness of such servo valve and its speed

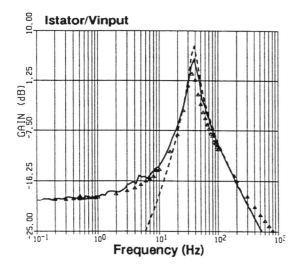

Fig. 9. Servo valve (pseudo)Bode gain curves.

capabilities.

4. SERVO VALVE MODELING AND IDENTIFICATION.

Bond-Graphs are a pictorial representation of dynamic systems. Using few symbols, this technique provides a model of any system combining many interacting energy domains. Each symbol corresponds to a physical element and can be obtained easily. Moreover, the deriving of state equations from the Bond-Graphs are easy to obtain and then can be introduced in any simulation software (ACSL for instance).

The servo valve combines several energy domains and components: electronic one for the unit control, electrical and magnetic ones for the motor, mechanical one for the rod and hydraulic one for the spool valve. Fig (8) shows the general Bond-Graph model obtained.

The motor is assimilated to a modulated gyrator producing a torque on its rotor axis. The rotor is characterized by its inertia J and viscous friction B. A modulated transformer converts rotation into translation. The dynamic of the spool valve is determined by its mass, its viscous friction and by the action of a spring designed to bring back the motor to the zero position.

The spool valve position modulates two hydraulic resistances which determine the relation between the pressure and the load flow.

On the other hand the angle position of the rotor is measured , compared to an angle position control input. The error between these two signals is used to control the stator current of the motor.

The fig (9) shows the two main gain curves (respectively the position gain and the current gain), obtained from the linear and nonlinear model compared to experimental results made with the servo valve without hydraulic load. In the linear case, these gains derive from the transfer functions (Bode gains) whereas in the non-linear case they stem from the first harmonic signal of each variable (pseudoBode gains).

The servo valve bandwidth is about 45 Hz , with a current resonance at approximately 40 Hz. The non-linear model is more precise than the linear one especially for the stator current in low frequencies.

5. MODELING AND SIMULATION OF THE SUSPENSION TEST RIG.

The Bond-Graph model is represented on fig (10) The hydropneumatic gas spring and the hydraulic damper are non linear. The main difficulties were to model the flow pipe to take into account the viscous

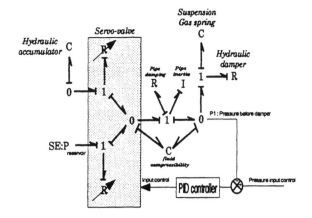

Fig. 10. Suspension test rig Bond-Graph model.

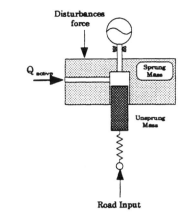

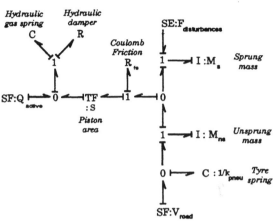

Fig. 12. 1/4 hydropneumatic active suspension Bond-Graph model.

damping and inertia. To avoid causality problems, a double port C has been introduced to take into consideration not only the oil compressibility in the chamber before damper but also in the pipe.

Comparing to the real experience, the simulation (fig (11)) shows the same variations for state variables. P_1 and its control reference are identical, P_2 the pressure after damper grows slowly, creating a flow through the hydraulic damper. This flow is compensated by the flow delivered by the servo valve to maintain P_1 constant. The spool valve tends to come back to the zero position, limiting the flow as P_2 becomes constant.

Thanks to this simulation, it can be concluded that the suspension test rig model is correct by reproducing the same phenomena as experimental results.

Moreover, this adequacy between simulation and experience demonstrates the validity of the servo valve model by its use in a specific application such as this test rig.

6. EXTENSION TO A ¼ HYDROPNEUMATIC ACTIVE SUSPENSION.

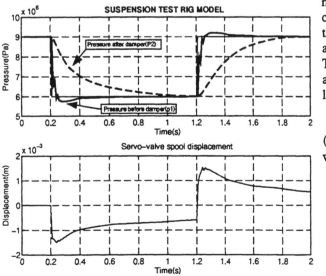

Fig. 11. Suspension test rig simulation.

To evaluate the performances of the actuator, its model has been added to a ¼ hydraulic suspension Bond-Graph model represented fig (12). This model takes into consideration not only the non linearities of the hydropneumatic spring C and the hydraulic damper R but also the piston Coulomb friction. This one has been modeled by the ' Reset Integrator Model ' (Haessig, 1991).

To test the servo valve, a disturbance force step is applied on sprung mass to simulate a turning maneuver , an acceleration or a braking. The control system purpose consists on maintaining the sprung mass in the same position despite the action of disturbances force on it (Karnopp, 1987). To reach this objective, a flow controlled by an actuator acts in the piston chamber following the law :

$$Q = f(g\ Z_{rel} + h V_{rel})$$

(1)
with:

Z_{rel} : The relative displacement between sprung and unsprung mass (suspension deflection)

V_{rel} : The relative velocity between sprung and unsprung mass (suspension velocity)

Q : the active flow

f(x): the actuator dynamic function
(f(x)=1 in the ideal active system.)
g,h : the feedback gains

Fig (13) compares the passive system, the ideal active system, and the real active system taking into account the servo valve dynamic. While the sprung mass gets down of 8cm in the passive case, the sprung mass only gets down of about 2cm to reach a final position less than 0.5 cm in the active case. This final position corresponds to the additional tire distortion produced by disturbances forces. Results obtained with or without the actuator dynamic are almost identical demonstrating the large potentiality of such actuator.

7. CONCLUSION.

By evaluating its behavior for several automotive applications, we can conclude to the high potentiality of this new direct-drive servo valve coming from its high speed and simple architecture.

The servo valve modeling using Bond-Graphs technique enables a simple insertion of that model in automotive suspension Bond-Graph models. The comparison between experiences and simulations testifies to the efficiency of Bond-Graph technique.

Moreover, with the actuator Bond-Graph model, any other applications can be studied. By evaluating the influence of each constitutive elements of the system, each control parameter, one can find an optimal design.

Finally, the high performances of this direct-drive servo valve adding to the use of powerful simulation Bond-Graph tools enables the implementation of many others applications requiring speed, power, accuracy, and especially low cost.

REFERENCES.

Borne, P., Dauphin-Tanguy, G et al (1992). Modélisation des systèmes physiques par Bond-graphs. In: *Modélisation et identification des processus*, (Technip), **Tome 2**, Chap 5, pp. 25-80.

Faisandier, J (1986). Electronique associée à l'hydraulique. Les nouveaux amplificateurs opérationnel et les nouveaux aimants peuvent révolutionner les servo valves. *Congrès SIA d'Angers 'l'hydraulique et l'automobile'* pp72-79

Guillon, M. (1991a). Intérêt de la transmission hydraulique. *Techniques de l'ingénieur.* {**B5 IV**}, B 6070.

Guillon, M. (1991b). Asservissements hydrauliques et électrohydrauliques. *Techniques de l'ingénieur.* {**B5 IV**}, B 6071.

Haessig, D.A. , Friedland, Jr.B. (1991). On the modeling and simulation of friction. *ASME Journal of Dynamic Systems, Measurement and Control*, **Vol 113**, pp 354-362.

Karnopp, D. (1987). Active Suspensions Based on Fast Load Levelers. *Veh.Sys.Dyn.***Vol 116** pp 355-380.

Menard, C. (1994). A new multipurpose servo valve for fast applications. *SAE Congress-Detroit*.

Merrit, H.E. (1967). *Hydraulic Control System*, (John Willey & Sons).

Mitchell and Gauthier Associates. (1990). *ACSL Manual*, Concord, Mass.

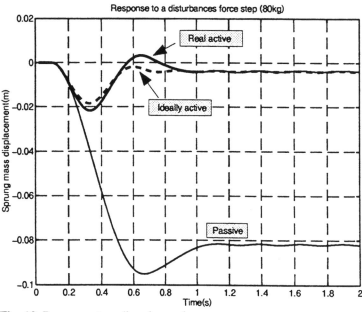

Fig. 13. Response to a disturbance force step.

MATHEMATICAL MODELING OF AUTOMOTIVE THREE WAY CATALYTIC CONVERTERS WITH OXYGEN STORAGE CAPACITY

F. Aimard*, S. Li**, M. Sorine**

*Renault, Direction de la Recherche, 9-11 avenue du 18-Juin-1940 92500 Rueil Malmaison France
**INRIA, 78153 Le Chesnay Cedex, France, Michel.Sorine@inria.fr

Abstract. Mathematical models of monolithic three way catalytic converters are developed which account for transient conversion efficiency and oxygen storage capacity. These models, with different degrees of complexity, are intended for use during several stages of the design of A/F ratio controllers and On Board Diagnostic systems for SI-engines. First a one space-dimensional model, accounting for thermal effects, is presented. Then two lumped parameter models are obtained, in the case of the hot phase, with 5 and 3 state variables respectively for use in long-term simulations or real-time computations.

Key Words. Pollution reduction; three way catalytic converter; oxygen storage capacity; On Board Diagnosis; Fuel/Air Ratio control

1. INTRODUCTION

Today stringent exhaust emission legislations for new cars have resulted in a wide use of Three Way Converter (TWC) for Spark-Ignition engines. Exhaust gases pass through a TWC to reduce pollution effects: residual hydrocarbons (HC) and carbon monoxyde (CO) are oxidized and nitrogen oxides (NO_x) are reduced. Catalyst precious metals are used to promote oxidation/reduction activity. High efficiencies of TWCs (more than 90 percent) are obtained within a narrow range of Fuel/Air Ratio (FAR) near the stoichiometric FAR. This narrow range is achieved by using an intake FAR closed-loop controller based on an Exhaust-Gas-Oxygen (EGO) sensor placed before the TWC (Hamburg and Shulman (1980), Aström and Wittenmark (1989)). Standard EGO sensors have a relay-like response, the stoichiometric FAR being the switching point (Baker and Verbrugge (1994)). With this nonlinear characteristic, closed-loop FAR limit cycles with tightly controlled mean value, amplitude and period, are normal operating conditions for TWCs (Katashiba et al. (1991), Falk and Mooney (1980)). Legislation in the USA (e.g. California On-Board-Diagnostic II since 1994), and in Europe in a next future, demands increased efficiency and real-time conversion efficiency monitoring. This leads to sophisticated control and diagnostic systems. For example dual EGO sensors methods use a second EGO sensor downstream of the TWC to monitor input/output oxygen balance (Koupal et al. (1991)). During normal activity, the output FAR limit cycle of a new TWC is reduced and de-layed, as would be the output of a nonlinear low-pass filter. For an aged TWC, the output tends to fluctuate as the input: dual EGO sensors methods are based on this input/output comparison. The physical phenomena involved in this input/output behavior is the Oxygen Storage Capacity (OSC) of Ceria which is added to the washcoat of today monolithic TWCs. The importance of OSC comes from its close relation to CO/NO_x conversion efficiency: OSC estimation is a way to monitor TWC efficiency (Fisher et al. (1993), Hepburn and Gandhi (1992), Meitzler (1980)).

Published mathematical models of automotive TWCs have been developed mainly for a better understanding of the underlying physical and chemical phenomena, and to improve TWC design (Byrne and Norbury (1993), Herz (1981), Heck et al. (1976), Heck et al. (1976), Young and Finlayson (1976a), (1976b)). A particular attention has been paid to the analysis of transient thermal and conversion characteristics (Oh and Cavendish (1982), Chen et al. (1988), Please et al. (1994)) and to the numerical analysis of these distributed parameters equations (Young and Finlayson (1976a) and Varma et al. (1976), Dochain et al. (1992), for more general chemical reactors), the dominant approach being the orthogonal collocation approximation.

The above models have severe drawbacks when the objective is to design and simulate new FAR controllers or OBD systems: they are too complex for long term simulations and do not take OSC into account. The purpose of this paper is twofold: A Partial Differential Equation (PDE) model of TWC efficiencies and OSC, described via ceria activity, is proposed (Section 2). It can be used for accurate performances simulations, and to a less extent, to TWC design. Simplified models, in fact Ordinary Differential Equations (ODE), are then derived that still describe conversion efficiencies and OSC (Section 3). Comparisons between simulations and experimental data are presented (Section 4).

Notation

- State variables:
- $T_g(x,t)$, $T_s(x,t)$: gas and converter surface temperatures,(K).
- $C_i(x,t)$: species i concentration in the gas stream, i=HC, O_2, CO, NO_x, (mole/m^3).
- $\theta(x,t)$: fraction of OSC filled with O_2.

- Parameters:
- ϵ: void fraction of the monolith (V_{void}/V)
- ρ_g, ρ_s: gas and solid densities (Kg/m^3).
- C_{pg},C_{ps}: gas, solid specific heats $(J/(Kg.K))$.
- K_s: solid thermal conductivity $(J/(m.s.K))$.
- h: heat transfer coefficient between gas and solid $(J/(s.K.m^2))$.
- S: catalyst surface in contact with gas per unit of reactor volume (m^{-1}).
- ΔH: line vector of reaction heat releases $(J/mole)$.
- N_s: number of active sites (precious metals) per unit of reactor volume $(mole/m^3)$.
- O_{sc}: oxygen storage capacity on Ce sites per unit of reactor volume $(mole/m^3)$.
- ξ_1, ξ_2: average numbers of carbon and hydrogen atoms in exhaust hydrocarbons.
- ξ_3: average number of oxygen atoms in exhaust nitrogen oxides, also called NO_x.
- k_{inhib}: inhibition coefficient of NO_x reduction in presence of oxygen.
- $k_{j,0}$, E_j: coefficient and activation energy of reaction j.
- R_g: gas constant.
- Q_g: mass flow rate of the gas (Kg/s).
- V, L : converter volume (m^3), length (m).
- T_{air}, $C_{O_2}^{air}$: exterior air temperature and oxygen concentration.

2. PDE MODELS OF TWCs

Chemical reactions and kinetics. Many reactions occur inside a TWC, but it results from comparisons with experimental data, that a small number of reactions can accurately represent mass and thermal energy balances and conversion rates. The considered reactions have been chosen on a trial-and-error basis, until a good accordance with measurements was achieved. The result is then probably dependent upon the experimental setting, described later. For example,

as H_2 was not measured and has a behavior similar to that of CO, these two species are aggregated into an equivalent one, CO_{eq}. A satisfactory reaction mechanism is the following simple system:

$$
\begin{array}{rcl}
CO_{eq} + \frac{1}{2}O_2 & \rightarrow & CO_2 \text{ (or } H_2O) \\
C_{\xi_1}H_{\xi_2} + (\xi_1 + \frac{\xi_2}{4})O_2 & \rightarrow & \xi_1 CO_2 + \frac{\xi_2}{2}H_2O \\
\frac{1}{\xi_3}NO_{\xi_3} + CO_{eq} & \rightarrow & CO_2 \text{ (or } H_2O) \\
& & + \frac{1}{2\xi_3}N_2 \\
O_2 + 2Ce_2O_3 & \rightarrow & 4CeO_2 \\
CO_{eq} + 2CeO_2 & \rightarrow & CO_2 \text{ (or } H_2O) \\
& & + Ce_2O_3
\end{array}
$$

The three first reactions describe the Three Way Conversions taking place on the precious metals. Oxygen storage and release on ceria, are described, following e.g. Fisher *et al.* (1993) by the two last reactions. The reaction rates follow standard Arrhenius equations, except for the NO_x reduction where a denominator is introduced to modelize inhibition of this reduction when O_2 is in excess (the same kind of inhibition model is used in Oh and Cavendish (1982), Chen *et al.* (1988) for oxidation). This phenomenon is important to explain the effect of OSC on NO_x conversion efficiency. The following matrix notations will be used to describe the concentration vector C, the reaction rate vector A and the matrices of yield coefficients K_C, K_θ:

$$
C = \begin{pmatrix} C_{HC} & C_{O_2} & C_{CO_{eq}} & C_{NO_x} \end{pmatrix}^T
$$

$$
A(C,\theta) = \begin{pmatrix} k_1 C_{CO_{eq}}\sqrt{C_{O_2}} \\ k_2 C_{HC}\sqrt{C_{O_2}} \\ k_3 \dfrac{C_{CO_{eq}}C_{NO_x}}{1 + k_{inhib}C_{O_2}} \\ k_4 C_{O_2}^{\alpha_{O_2}}(O_{sc}(1-\theta))^{\beta_{O_2}} \\ k_5 C_{CO_{eq}}^{\alpha_{CO}}(O_{sc}\theta)^{\beta_{CO}} \end{pmatrix}
$$

$$
K_C = -\frac{N_s}{\epsilon}\begin{pmatrix} 0 & 1 & 0 & 0 & 0 \\ \frac{1}{2} & \xi_1 + \frac{\xi_2}{4} & 0 & 1 & 0 \\ 1 & 0 & 1 & 0 & 1 \\ 0 & 0 & \frac{1}{\xi_3} & 0 & 0 \end{pmatrix}
$$

$$
K_\theta = \frac{N_s}{O_{sc}}\begin{pmatrix} 0 & 0 & 0 & 2 & -1 \end{pmatrix}
$$

with $k_j = k_{j,0}\exp\left(\dfrac{-E_j}{R_g T_s}\right)$ for $j = 1,\ldots,5$.

Conservation of mass and energy. A transient, one-dimensional model is chosen. Following Oh and Cavendish (1982), radial variations of temperature and concentrations are neglected. All diffusions are neglected, except the axial one for temperature in the solid. Reactions are assumed to occur only on the external surface of the catalytic wall, so that we consider only concentrations in the gas, and on ceria for O_2. The gas velocity is given by $V_g = \dfrac{Q_g L}{\epsilon \rho_g V}$. The basic distributed models can now be stated.

One-dimensional model with heat transfer.

$$\frac{\partial T_g}{\partial t} = -V_g \frac{\partial T_g}{\partial x} + \frac{Sh(T_s - T_g)}{\varepsilon \rho_g C_{pg}}$$

$$\frac{\partial T_s}{\partial t} = \frac{K_s}{\rho_s C_{ps}} \frac{\partial^2 T_s}{\partial x^2} + \frac{Sh(T_g - T_s)}{(1-\varepsilon)\rho_s C_{ps}}$$

$$- \frac{\Delta H}{(1-\varepsilon)\rho_s C_{ps}} A(C,\theta)$$

$$\frac{\partial C}{\partial t} = -V_g \frac{\partial C}{\partial x} + K_C A(C,\theta)$$

$$\frac{\partial \theta}{\partial t} = K_\theta A(C,\theta) \qquad (1)$$

Appropriate boundary and initial conditions are:

$$T_g(0,t) = T_g^{in}(t) \quad \frac{\partial T_s}{\partial x}(0,t) = \frac{\partial T_s}{\partial x}(L,t) = 0$$

$$T_g(x,0) = T_{air} \quad C(0,t) = C^{in}(t) \qquad (2)$$

$$C(x,0) = C^{air}(x) \quad \theta(x,o) = 1$$

C^{in} and C^{air} are the vectors of input concentrations and of species concentrations in the air.

One-dimensional model with constant temperature. $T_s = T_g$ constant is the important case of a warm converter. The model depends upon the operating point Q_g, T_g, through V_g and k_i:

$$\frac{\partial C}{\partial t} = -V_g \frac{\partial C}{\partial x} + K_C A(C,\theta)$$

$$\frac{\partial \theta}{\partial t} = K_\theta A(C,\theta) \qquad (3)$$

The boundary and initial conditions reduce to:

$$C(0,t) = C^{in}(t) \qquad C(x,0) = C^{air}(x)$$

$$\theta(x,0) = 1 \qquad (4)$$

Similar models are used for bioreactors (Bastin and Dochain (1990)).

The Normalized Fuel/Air Ratio (NFAR). λ is defined as[1]:

$$\lambda = \frac{(\text{mass fuel / mass air})}{(\text{mass fuel / mass air})_{stoichiometry}}$$

λ can be computed from engine exhaust gas composition (Fukul *et al.* (1989), Abida and Claude (1994)). For example, for the fuel at hand, the following "Five gases formulae" holds:

$$\lambda = 1 +$$

$$A_z \frac{\frac{1}{2}Y_{CO_{eq}} - Y_{O_2} - \frac{\xi_3}{2}Y_{NO_x} + (\xi_1 + \frac{\xi_2}{4})Y_{HC}}{0.99 - Y_{HC} - Y_{O_2} - Y_{CO_{eq}} - Y_{CO_2} - \frac{1}{2}Y_{NO_x}}$$

where air composition is $O_2 + A_z N_2$ and Y_i is the mole fraction of species i. Experiments have shown that the denominator of this expression is constant over a large range of operating condi-

---

[1] Often, λ denotes the inverse: the Normalized Air/Fuel ratio

tions (Aimard (1995)), so that:

$$\lambda(x,t) = 1 + K_\lambda C(x,t) \text{ with}$$

$$K_\lambda = k_\lambda \left(\begin{array}{cccc} \xi_1 + \frac{\xi_2}{4} & -1 & \frac{1}{2} & -\frac{\xi_3}{2} \end{array} \right) \qquad (5)$$

As it can be checked, $K_\theta = \frac{2\varepsilon}{k_\lambda O_{sc}} K_\lambda K_C$, so that, combining this identity with (1) leads to the following interesting relation between the NFAR inside the converter and θ:

$$\frac{\partial \lambda}{\partial t} = -V_g \frac{\partial \lambda}{\partial x} + \frac{k_\lambda O_{sc}}{2\varepsilon} \frac{\partial \theta}{\partial t}$$

$$\lambda(0,t) = \lambda^{in}(t), \quad \lambda(x,0) = 0 \qquad (6)$$

$$\theta(x,0) = 1$$

This relation is useful for OSC identification.

3. ODE MODELS OF TWCs

Model simplification by space discretization. A frequently used spatial approximation for models like (1) and (3) is the orthogonal collocation method, efficient for model reduction (Dochain *et al.* (1992), Varma *et al.* (1976), Young and Finlayson (1976a), Li (1995)). In fact, even a lumped model using only two points, at input and output, gives already good results when the parameters are identified after discretization. This model, basis for further reduction, is given for the case of equations (3), (4). $C(t)$ is again the vector of output concentrations:

$$\frac{dC}{dt} = -\frac{V_g}{L}(C - C^{in}) + K_C A(C,\theta)$$

$$\frac{d\theta}{dt} = K_\theta A(C,\theta) \qquad (7)$$

A model of comparable complexity is used in Katashiba *et al.* (1991). Multiplying (7) by K_λ leads to a relation similar to (6):

$$\frac{d\lambda}{dt} = -\frac{V_g}{L}(\lambda - \lambda^{in}) + \frac{k_\lambda O_{sc}}{2\varepsilon} \frac{d\theta}{dt}$$

$$\lambda(0) = 0, \quad \theta(0) = 1 \qquad (8)$$

Model simplification by reactant aggregation. This further simplification of (7) is based on a chemical approach: CO and $H_{\xi_1}C_{\xi_2}$ are aggregated into an "equivalent" reducing species Red, O_2 and NO_{ξ_3} are aggregated into an "equivalent" oxidizing species Ox. More precisely, defining:

$$\tilde{C} = \left(\begin{array}{c} C_{Red} \\ C_{Ox} \end{array} \right) = \tilde{K} C \text{ with}$$

$$\tilde{K} = \left(\begin{array}{cccc} 2(\xi_1 + \frac{\xi_2}{4}) & 0 & 1 & 0 \\ 0 & 1 & 0 & \frac{\xi_3}{2} \end{array} \right) \qquad (9)$$

it is easily checked that NFAR is a function of $\tilde{C}$:

$$\lambda = 1 + \tilde{K}_\lambda \tilde{C} \text{ with } \tilde{K}_\lambda = k_\lambda \left(\begin{array}{cc} \frac{1}{2} & -1 \end{array} \right) \qquad (10)$$

An equation for $\tilde{C}$ can now be obtained. Multiplying (7) by $\tilde{K}$ is not sufficient as $\tilde{K}K_C A(C,\theta)$ is not a function of $\tilde{C}$ and θ. So, a new reaction mechanism is postulated, where Σ represents sites with OSC:

$$
\begin{aligned}
Red + \tfrac{1}{2}O_x &\rightarrow CO_2 \text{ (or } H_2O) \\
Ox + 2\Sigma &\rightarrow 2\Sigma O \\
Red + \Sigma O &\rightarrow CO_2 \text{ (or } H_2O)
\end{aligned}
$$

The kinetics being now defined by:

$$
\begin{aligned}
\tilde{A}(\tilde{C},\theta) &= \begin{pmatrix} k_I C_{Red}\sqrt{C_{Ox}} \\ k_{II} C_{Ox}^{\alpha_{Ox}} (O_{sc}(1-\theta))^{\beta_{Ox}} \\ k_{III} C_{Red}^{\alpha_{Red}} (O_{sc}\theta)^{\beta_{Red}} \end{pmatrix} \\
\tilde{K}_{\tilde{C}} &= -\frac{N_s}{\varepsilon} \begin{pmatrix} 1 & 0 & 1 \\ \frac{1}{2} & 1 & 0 \end{pmatrix} \\
\tilde{K}_{\theta} &= \frac{N_s}{O_{sc}} \begin{pmatrix} 0 & 2 & -1 \end{pmatrix}
\end{aligned}
$$

with $k_j = k_{j,0} \exp\left(\frac{-E_j}{R_g T_s}\right)$ for $j = I, II, III$. An equation similar to (7), based on this reactional model, with only three state variables has been successfully identified from experiments:

$$
\begin{aligned}
\frac{d\tilde{C}}{dt} &= -\frac{V_g}{L}(\tilde{C} - \tilde{C}^{in}) + \tilde{K}_{\tilde{C}}\tilde{A}(\tilde{C},\theta) \\
\frac{d\theta}{dt} &= \tilde{K}_{\theta}\tilde{A}(\tilde{C},\theta)
\end{aligned}
\quad (11)
$$

(10) still holds as, again, $\tilde{K}_{\theta} = \frac{2\varepsilon}{k_\lambda O_{sc}}\tilde{K}_\lambda \tilde{K}_C$. These models are convenient for long term simulations or real time applications.

4. EXPERIMENTAL RESULTS

The above models have been identified and used in several applications. A TWC with 2 cordierite monoliths of $1.8dm^3$ and 400 cell/$inch^2$ has been used for experiments. It has a classical impregnation from Johnson Matthey of $50g/ft^3$ precious metal amount, with the ratio 5/0/1 of (Pt, Pd, Rh). The engine used was a Renault J7TL740, a 2,2 litres with 4 cylinders. Pollutants measures come from 3 kinds of gas analyzers: a 2 channels flame ionization detector for HC measurements (time response of 1ms and time delay of 10ms), a chemi-luminescence analyser for NO_x (time response of 100ms and time delay of 100ms), a non dispersive infrared spectrometer for CO, CO_2 (time response of 30ms and time delay of 100ms). The experimental car, a Renault 21 Nevada TXE, besides its serial FAR controller (EGO sensor, 4 fuel injectors, a Fenix computer), was equipped with proportional oxygen sensors for Normalized Fuel Air Ratio (NFAR) measurements at the outlet of the engine and before and after the catalyst, and thermocouple sensors for gas temperature before and after the catalyst (Aimard (1995)). Examples of long term simulations are shown in Figures 1 and 2 where the bottom curve is the speed profile along the phase 1 of an ECE cycle (cold start, then maxima at 15;

30; 50 km/h). TWC Light-off can be seen after $60s$. Figures 3 and 4 show some short term simulations: the influence of input NFAR frequency on output NFAR and CO emissions.

A simulator comprising an engine model and the

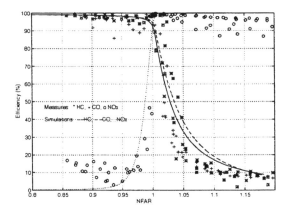

Fig. 1. Measured and computed TWC static efficiencies

TWC model (7) is currently used at Renault and Inria to design FAR controllers and OBD systems (Aimard et al. (1995)). In that case N_s and O_{sc} are used to modelize aged TWCs.

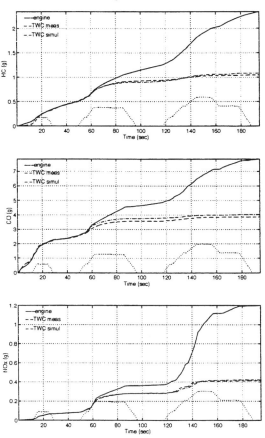

Fig. 2. Cumulated emissions along phase 1 of an ECE cycle: measured at engine output, measured and computed at TWC output

5. CONCLUSION

Several mathematical models of automotive three way catalytic converters have been presented. A relatively complex distributed model allows to study converter efficiency during thermal and concentration transients. A model of medium complexity can be identified to study HC, CO and NO_x efficiencies and their relations with oxygen storage capacity during concentration transients. Coupled with an engine pollution model, it allows long term simulations (e.g. complete ECE cycle) of new or aged converters and this is a key feature for the design of Fuel/Air ratio controllers and OBD systems. A low complexity model can also be identified and gives an account of the relation between oxygen storage and Fuel/Air ratio. This is useful in real time control or diagnosis applications.

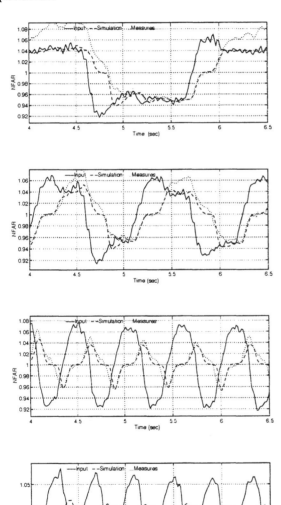

Fig. 3. NFAR: measured at TWC input, measured and computed at TWC output

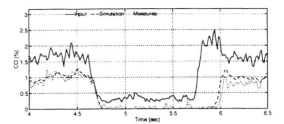

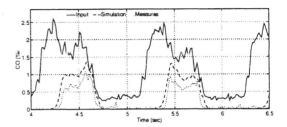

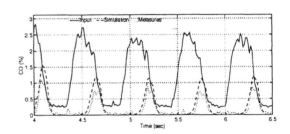

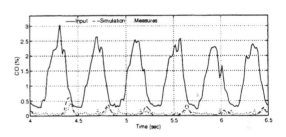

Fig. 4. CO: measured at TWC input, measured and computed at TWC output

6. REFERENCES

Abida, J. and D. Claude (1994). Spark ignition engine and pollution emission: new approaches in modelling and control. *Int. J. of Vehicle Design* **15**(3/4/5), 494–508.

Aimard, F. (1995). Modélisation du comportement dynamique du pot catalytique. Utilisation en contrôle moteur et diagnostic catalyseur. Nouveau doctorat. Université de Paris–Dauphine.

Aimard, F., C. Cussenot, S. Li and M. Sorine (1995). Modélisation d'un moteur thermique dépollué en vue de la conception de systèmes de commande et de surveillance/diagnostic. In: *Journées Automatique et Automobile* (A. Oustaloup, Ed.). LAP, Université Bordeaux I.

Aström, K.J. and B. Wittenmark (1989). *Adaptive control*. Addison-Wesley.

Baker, D.R. and M.W. Verbrugge (1994). Mathematical analysis of potentiometric oxygen sensors for combustion-gas streams. *AIChE Journal* **40**(9), 1498–1514.

Bastin, G. and D. Dochain (1990). *On-line estimation and adaptive control of bioreactors.* Elsevier, Amsterdam.

Byrne, H. and J. Norbury (1993). Mathematical modelling of catalytic converters. *Math. Engng. Ind.* **4**(1), 27–48.

Chen, D.K.S., E.J. Bisset, S.H. Oh and D.L. Van Ostrom (1988). A three-dimensional model for the analysis of transient thermal and conversion characteristics of monolithic catalytic converters. number 880282 In: *SAE Technical paper series.* SAE.

Dochain, D., J.P. Babary and N. Tali-Maamar (1992). Modelling and adaptive control of nonlinear distributed parameter bioreactors via orthogonal collocation. *Automatica* **28**(5), 873–883.

Falk, C.D. and J.J. Mooney (1980). Three-way conversion catalysts: effect of closed-loop feedback control and other parameters on catalyst efficiency. number 800462 In: *SAE Technical paper series.* SAE.

Fisher, G.B., J.R. Theis, M.V. Casarella and S.T. Mahan (1993). The role of ceria in automotive exhaust catalysis and OBD-II catalyst monitoring. number 931034 In: *SAE Technical paper series.* SAE.

Fukui, T., Y. Tamura, S. Omori and S. Saitoh (1989). Accuracy of A/F calculation from exhaust gas composition of SI engines. number 891971 In: *SAE Technical paper series.* SAE.

Hamburg, D.R. and M.A. Shulman (1980). A closed loop A/F control model for internal combustion engines. number 800826 In: *SAE Technical paper series.* SAE.

Heck, R.H., J. Wei and J.R. Katzer (1976). Mathematical modeling of monolithic catalysts. *AIChE Journal* **22**(3), 447–484.

Hepburn, J.S. and H.S. Gandhi (1992). The relationship between catalyst hydrocarbon conversion efficiency and oxygen storage capacity. number 920831 In: *SAE Technical paper series.* SAE.

Herz, R.K. (1981). Dynamic behavior of automative catalysts. Catalyst oxidation and reduction. *Ind. Eng. Chem. Prod. Res. Dev.* (20), 451–457.

Katashiba, H., M. Nishida, S. Washino, A. Takahashi, T. Hashimoto and M. Miyake (1991). Fuel injection control systems that improve three way catalyst conversion efficiency. number 910390 In: *SAE Technical paper series.* SAE.

Koupal, J.W., M.A. Sabourin and W.B. Clemmens (1991). Detection of catalyst failure on-vehicle using the dual oxygen sensor method. number 910561 In: *SAE Technical paper series.* SAE.

Li, S. (1995). Modélisation de pots catalytiques et de sondes de richesse. Application à la dépollution des moteurs thermiques. Nouveau doctorat. Université de Paris–Dauphine.

Meitzler, A.H. (1980). Application of exhaust-gas-oxygen sensors to the study of storage effects in automotive three-way catalysts. number 800019 In: *SAE Technical paper series.* SAE.

Oh, S.H. and J.C. Cavendish (1982). Transients of monolithic catalytic converters: response to step changes in feedstream temperature as related to controlling automobile emissions. *Ind. Eng. Chem. Prod. Res. Dev.*

Please, C.P., P.S. Hagan and D.W. Schwendeman (1994). Light-off behavior of catalytic converters. *SIAM J. Appl. Math.* **54**(1), 72–92.

Varma, A., C. Georgakis, N.R. Amundson and R. Aris (1976). Computational methods for the tubular chemical reactor. *Computer methods in applied mechanics and engineering* **8**, 319–330.

Young, L.C. and B.A. Finlayson (1976a). Mathematical models of the monolith catalytic converter: Part I: Development of model and application of orthogonal collocation. *AIChE Journal* **22**(2), 331–343.

Young, L.C. and B.A. Finlayson (1976b). Mathematical models of the monolith catalytic converter: Part II: Application to automobile exhaust. *AIChE Journal* **22**(2), 343–353.

EMERGENT TECHNIQUES FOR MOBILE ROBOTS AND AUTONOMOUS VEHICLES

Giovanni Ulivi *

Dip. di Meccanica e Automatica - III Università di Roma - ulivi@labrob.ing.uniroma1.it

Abstract. New control methodologies (fuzzy, neural, genetic) are often used to design the control systems of mobile robots. This paper analyzes several significant cases in the fields of feedback control, obstacle avoidance, and automotive components and tries to arrive to some general conclusions.

Key Words. Mobile robots, autonomous vehicles, fuzzy control, neural networks, genetic algorithms

1. INTRODUCTION

Mobile robotics is receiving more and more attention as demonstrated by the number of dedicated conferences and the existence of active technical committees in international associations[1].

This interest has two aspects. An autonomous mobile agent can be useful in many applications, ranging from space exploration to less esoteric mail distribution services in offices or spraying in cultivated fields. Less diffusive but noteworthy uses can be in the field of dangerous areas surveillance (nuclear and chemical plants, caves) and sick or disabled people assistance (Cox and Wilfong, 1990).

Moreover a mobile robot is a rather simple yet versatile benchmark, where researchers can test quite conventional algorithms as simple as PID position controllers in the servos and as complex as intelligent planners which determine the best path in a cluttered environment. Sensing is another very important issue. Different kinds of sensors are used, and many processing techniques can be used to obtain information about the external environment.

From a control theory point of view, a mobile robot presents very interesting peculiarities, being a nonlinear, nonholonomic system. If we are interested to steer the vehicle from one point to another (with a prescribed orientation) and we are not concerned with obstacles, we can reasonably decide to find a control law which renders the target position attractive. Even in this simple case, complex feedback laws are necessary, due to the nature of the system. This difficulties motivated

the study of several kinds of rule-based controllers. The result is an on-going, very interesting debate between the sustainers of "classical" and "intelligent" controllers, which is documented by many papers. A short outline of these problems will be given in section 2.

When an actual mobile robot is considered, many constraints come into play. Obstacles, steering limitations, odometry errors are factors that cannot be taken all into account by simple controllers, as the ones described before and a complex control structure, composed by several tasks, is necessary.

According with the principle of increasing complexity–decreasing precision, "intelligent" controllers find more space and motivations where the complexity of the controlled system is higher. For space reasons, here we will consider only the obstacle avoidance problem with some extensions to sensing and navigation. This problem is the subject of section 3.

More interesting, from an industrial point of view, are automotive applications. Many important car builders are doing research to produce new intelligent car component which should make new cars more safe and comfortable, often using emergent technologies. Some of these researches are documented by published papers. Some representative cases are referred in section 4.

2. FEEDBACK CONTROL

Even considering an highly idealized behavior (i.e. neglecting slippage, frictions and so on), a mobile robot exhibits a very peculiar dynamic model. It is a nonlinear, underactuated and nonholonomic system (De Luca and Oriolo, to appear) which poses difficult control problems. Note that many other systems, important in the applications, can

---

[1] IFAC "Robotics" and "Intelligent Autonomous Vehicles" TC's, "IEEE Mobile Robot TC" of the Robotics and Automation Society

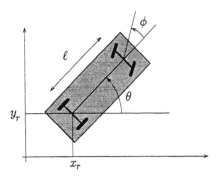

Fig. 1. Car-like robot

be described by such a model: car with trailers, space robots, underactuated manipulators.

Consider the model of a simple car-like robot (see figure 1):

$$
\begin{bmatrix} \dot{x}_r \\ \dot{y}_r \\ \dot{\theta} \\ \dot{\phi} \end{bmatrix} = \begin{bmatrix} \cos\theta \\ \sin\theta \\ \frac{1}{\ell}\tan\phi \\ 0 \end{bmatrix} u_1 + \begin{bmatrix} 0 \\ 0 \\ 0 \\ 1 \end{bmatrix} u_2, \quad (1)
$$

it has four state variables (generalized coordinates) $q = (x, y, \theta, \phi)$, where (x, y) are the Cartesian coordinates of the rear axle midpoint, θ is the angle between the car axis and the x axis of the reference frame and ϕ is the steering angle. There are two input variables, namely the driving velocity u_1 and the steering velocity u_2. As at least the first three coordinates are of interest to define the result of a manouvre, the system is clearly underactuated. Nonholonomy characterization requires complex mathematics and can be found in the cited reference; in a nutshell, it implies that vehicle can reach any point in the configuration space (x, y, θ) but it is not free to follow all the trajectories, a well-known consequence being the difficulty of parking a car in a narrow space.

As a consequence of these characteristics, a very strong result holds, valid for all nonholonomic systems:

No C^1 static feedback law exists which make a point q_e of the configuration space asymptotically stable.

Some remarks are in order:

- *Any* feedback law having those characteristics is ruled out, independently of the way used to design it.
- Dropping some of the hypothesis, a controller can be designed.
  - Open loop controllers can be easily found. In this case we obtain a *trajectory planner*, where u is independent of state variables and depends only on the starting and the ending points (see e.g. (Latombe, 1991)).
  - Using a nonsmooth (de Wit and Sordalen, 1992; Liu and Lewis, 1992) or time-varying feedback (Samson, 1993), closed loop controllers can be designed. They are more dif-

ficult to work out, but are less sensitive to modeling errors and disturbances.

- The result concerns *point* stabilization, i.e. in ideal conditions the vehicle reaches *exactly* the given target asymptotically. It does not apply when we can accept some error on the final position.

The typical fuzzy paradigm to design controllers is to extract human knowledge due to experience by experiments or interviews. Mobile robots are very well suited for this approach and this explains why so few papers can be found which use implicit knowledge (neural or genetic algorithms) for this problem. One of the best known examples of autonomous vehicle fuzzy control is given in (Kosko, 1992), where a set of 35 rules are used to back up a vehicle to a loading dock. In (Tanaka, 1994), a trailer is added and a more complex control problem is solved, again by using a fuzzy controller. An extension of the design paradigm, that has been applied to this problem, consists in the automatic generation of the rules from existing data (Wang and Mendel, 1992).

In all these works, an exact final position is not pursued. For example, trying with the program provided with the book by Kosko, one can easily verify that the vehicle's final position depends on the starting point and on controller parameters. This can be accepted in many practical cases, in particular if one takes into account all the other sources of error that may affect a real system. However it has to be considered also in performing a fair comparison between the two approaches.

A kind of intermediate approach is taken by other authors. In (Badreddin and Mansour, 1993),the gains of a standard PI controller are changed by a fuzzy supervisor during the trajectory execution. In this way the low-level controller satisfies the necessary conditions for stabilizability and a zero final error occurs. In a somewhat similar –and more complex– way, in (Deng and Brady, 1993), a fuzzy supervisor is superimposed to a feedback linearizing controller.

We can conclude that fuzzy controllers are much easier to design, mainly because they give poorer performances. However, in real word cases, their performances can be satisfying and moreover they can be augmented to tackle with a more complex word which includes obstacles, steering saturation, and so on. As a matter of fact, up to day, there is no way to include these important features in model based feedback controllers. In this category, only path planners, i.e. open loop controllers, can give useful results.

3. OBSTACLE AVOIDANCE

Avoiding obstacles is one of the most basic behaviors for an autonomous robot and it is the subject of many papers. The problem in its essence is very simple: "avoid an obstacle that (suddenly) appears in front of the robot" and requires very fast reactions. Despite its simplicity and its obviousness, the previous statement is quite fuzzy and indeed the problem is not so well posed as the one treated in the previous paragraph. Sensing obstacles adds uncertainty to the problem. As a matter of fact it is much easier to find papers tackling this problem by soft-computing approaches.

In (Borenstein and Koren, 1991) one of the best known algorithms for sensor-based obstacle avoidance is presented together with a thorough presentation of the problem and a survey of other existing methods. The system uses 24 ultrasonic sensors with a lobe width of 30°. The environment is represented by an occupancy grid with 10 by 10 cm square cells. The map is a square, the side of which measures 33 cells, and is centered on the robot. Even if the sensors have a large lobe of radiation, just the cells on the axes of the sensors are upgraded because, owing to the robot motion, all the cells are anyway sweeped. The upgrade law is quite heuristic, and no reference is made to probability theory. At each measurement, values in cells appearing empty are decremented and values in occupied cells are incremented, both by tunable values. The map is a matrix of about 1000 values (cells), representing the "likelihoods" of occupancy. All these data undergo a first "data reduction" which generates a "polar obstacle density": an histogram representing the dangerousness of 72 directions evenly distributed over a circle surrounding the robot. The histogram is then searched for "sufficiently wide valleys", the one closest to the planned direction is chosen to direct the robot motion. The algorithm runs on a 20 MHz 386 with a sampling time of 27 ms. Among the limitations, the author quote oscillations when following a wall (avoided by proper parameter tuning), cyclic behaviors and trapping in dead-end situations. The last two problems are overridden invoking a global path planner.

This solution is a prototype for many fuzzy obstacle avoidance algorithms; the same problem of oscillating behaviors and deadlocks is explicitly referred in (Reignier, 1994). As a matter of fact, all the heuristic solutions presented in the paper could be formulated in a fuzzy framework. Among the many works devoted to obstacle avoidance which employ fuzzy techniques, we can quote (Pin et al., 1992; Pin and Watanabe, 1993; Reignier, 1994; Demel et al., 1995). Other papers (Sugeno and Murakami, 1985; von Altrok et al., 1992; Braunstingl, 1995) embed obstacle avoidance in more general tasks, as parking a car, following a corridor at high speed or following a wall. Papers similar to (Borenstein and Koren, 1991) are (Braunstingl, 1995; Demel et al., 1995).

In the first, the readings from 12 ultrasonic sensors are used to build a "general perception vector" which points toward the most dangerous direction. A base of 33 rules controls the vehicle steering. In (Demel et al., 1995) an infrared laser scanner provides the measures to build a cartesian occupancy grid. The two nearest (a crisp concept!) cells at the right and at the left of the robot are used to determine the steering angle, using 149 rules. Among the cited works, this is the only which retains the use of a map where the measures are accumulated so as to reduce the effect of erroneous readings. About 200 rules are used in (von Altrok et al., 1992) to drive a model car at low speeds, whilst 600 rules are needed to take into account skidding.

These numbers can be compared with the 40 rule base inference running on the Z80 processor of the parking car of (Sugeno and Murakami, 1985). Incrementing the number of sensors and the number of linguistic labels produces a combinatorial growth of the rules and of the needed computing power, which in (von Altrok et al., 1992) is provided by a card with four transputers.

This problem is especially considered in (Reignier, 1994) and in (von Altrok et al., 1992). In the first paper the number of antecedents for each rule is partitioned in several rule banks. The results are then merged using suitable weighting factors. In the other a hierarchy of rules bases is proposed. Each base at the first level is specialized in revealing a driving situation which determines the behavior of the robot by an other level of rules. Another way to avoid high numbers of rules is proposed in (Nishimori and et al., 1994), where it is shown that a fuzzy-neural approach can be used to tune the rules so that a bank of just 7 rules can behave as well as a 49 rules one.

A different way to avoid the drawbacks of using many rules is used by the author in the obstacle avoidance system of a wheelchair for disabled people, which closely recalls (Borenstein and Koren, 1991). An overwiev of the project, which is somewhat similar to NAVCHAIR (Simpson et al., 1994; Bell et al., 1994), is given in (Ulivi, 1995)). Actually the use of rules is not strictly necessary to implement obstacle avoidance and the algorithm can be determined using more elementary concepts from fuzzy logic.

Figure 2 and 3 show a typical indoor environment together with the readings of 9 US sensors obtained from different locations (axes values are in centimeters). The locations, shown as circles, are 25 cm apart; the readings are reported by the arcs corresponding to the sensor lobe and have been obtained by an experimental set-up. In the first drawing the passage is sufficient to let the robot pass, in the other it is not.

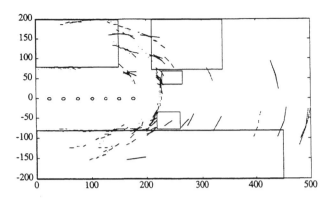

Fig. 2. Ultrasonic scene for a robot moving in an indoor environment. The passage is sufficient.

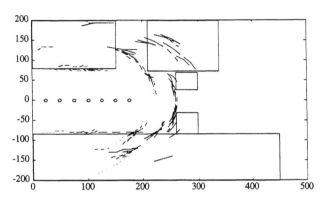

Fig. 3. Ultrasonic scene for a robot moving in an indoor environment. The passage is not sufficient.

In order to reduce the effects of sensor misreadings, a memory of past situations is highly effective. Instead of an instantaneous mapping from sensors to servos, a true evaluation of the risk of moving in a given direction can be performed using stored data. The main factor determining the risk is the distance d_i of a possible obstacle (segment) from the robot, which contributes to the risk density by its reciprocal k_d/d_i, being k_d a scaling factor. Moreover, older measures are less useful than new ones to determine the risk of moving along a given direction and odometry errors make them even misleading. Therefore, it is convenient to introduce a forgetting function, which weighs the previous perception by their age. Here a simple linear function is used, and, after four measurements (a 1 meter path), a reading is no more used. The risk density is finally computed by *and*-ing the risk values of the possible obstacles met in a given direction. Here, a 3 degree discretization of the front half-circle is used and the *and* function has been implemented by the *bounded-sum* operator.

Note that both the sensor and the movement spaces are intrinsically polar. This leads to the use of a polar representation of the environment, as shown in figures 4 and 5, where possible obstacles

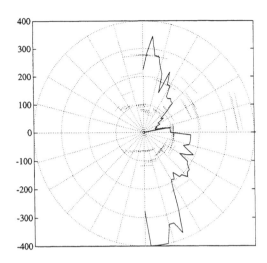

Fig. 4. Polar map of the environment showing the risk density after 3 measures.

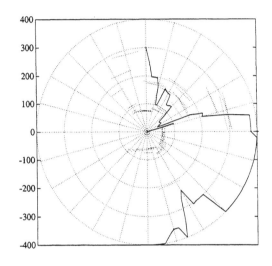

Fig. 5. Polar map of the environment showing the risk density after 7 measures.

(segments) and risk densities (continuous line) are drawn respectively at the third and the seventh measurement position. This representation does not require the time consuming coordinate transformations needed by a cartesian frame. Clearly, as the robot proceeds, the old segments undergo complex transformations on the polar representation. However, being the map local and the older measures less important, these transformations can be approximated by simpler ones.

Comparing the two diagrams, it can be easily seen that the scenario gets clearer when the robot approaches the passage. In this case, the opening was at the left of the robot and the risk density of figure 5 shows that it is maximally risky to proceed ahead or toward the right.

However, it remains to determine if the robot can traverse the passage, i.e. if it contains the robot. Assuming a normalization distance of 0.5 meters, the width of the robot is transformed in an angle

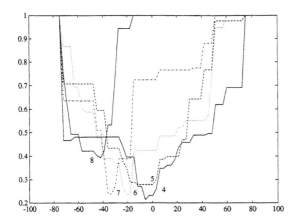

Fig. 6. Possibility for the robot to pass, shown at different positions. Passage at robot's left.

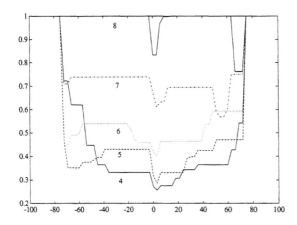

Fig. 7. Possibility for the robot to pass, shown at different positions. No passage.

(50 degrees) which, by using the Lukasiewitcz implication, is compared with the risk density. The results are given in figures 6 and 7 respectively for the passage at the left and for the case of figure 3, where no passage can be found. The numbers specify the measuring points.

This algorithm is very fast, in particular if, instead of keeping memory of the measures, just the risk density is recorded and modified step by step. This introduces some error, the effect of which is however negligible.

Obstacle avoidance is strictly connected with path following algorithms, see e.g. (Martinez *et al.*, 1993). Therefore two competitive behavior can appear at the same time, when an obstacle is detected. Sometimes the two behaviors are switched in a crisp way (Beom and Cho, 1995). This method can however provoke oscillatory phenomena. More appropriately for a fuzzy framework, they can be mixed in a smooth way as shown in (Saffiotti *et al.*, 1993; Yen and Pfluger, 1995). In the first paper, a desirability function is associated to each action to form a control scheme (e.g. *follow*). These functions, augmented with context information and with labels addressing the world objects they refer to, form the so-called Behav-

ior Schemas (e.g. *if there is no obstacle* (context) *then follow* (action) *a corridor* (object)). Behavior Schemas are the basis to determine the robot motion. They can be finalized to reach a target and blended in different ways, using typical fuzzy logic operators as negation, conjunction, and so on. This blending system for responsible of the smooth behavior of "Flakey", a mobile robot gaining the second-place in 1992 AAAI robot competition.

Sensor-based obstacle avoidance is a topic dominated by heuristics. Related human knowledge abounds and is expressed by simple rules. Clearly, this makes fuzzy methods very good candidates to generate high-performance algorithms. The main difficulty is in the combinatorial explosion of the number of rules and therefore of the needed computing power. A clever use of logic and/or a partition of the rule banks can help in alleviating this problem.

A completely different approach consists in the use of a neural network to build a direct mapping from the perception space (measures) to the action space (servos) from examples (Aguilar and Coutreras-Vidal, 1994) or just from a binary evaluation of the manouvre success (Sehad and Touzet, 1994). Neural network can tackle even with the problem of multiple reflections which affects ultrasonic devices, as shown in (Santos *et al.*, 1994). Some experimental results show a very good local mapping of the environment with 24 sensors. However, difficulties in sizing the inner layer are reported.

As the net learns directly from actual measures, there is no need to model the interactions between the sensors and the environment (a difficult task specially for ultrasonic units) or between the commands and the motion (the robot kinematical model). However, neural networks present two serious drawbacks which are the long training phase and the need of a complete redesign and training when a new sensor is added.

More strategical tasks, as map building or path planning, are not often addressed in the literature. In (Poloni *et al.*, 1995), a map is built from ultrasonic sensor readings by using simple fuzzy logic operators. The results are superior to those obtained by a probabilistic approach. In (Oriolo *et al.*, 1994), an A^* algorithm searches the fuzzy map to determine an optimal path. Considered cost functions include minimum risk and a weighted sum of risk and length. Genetic algorithms have been used for motion planning. In (Shibata and Fukuda, 1993), a fuzzy fitness function is used to evaluate the planned motion.

4. AUTOMOTIVE APPLICATIONS

New cars use more and more automatized components to improve safety (ABS, antiskid, airbag) and performance, both in term of comfort and pollution (electronic injection, adaptive suspensions, transmission trains, air conditioners). Intelligent vehicle highway systems (IHVS) are intensively studied and will require cars with a high degree of automation and complex sensory systems. Common characteristics of all these systems are the lack of exact requirements and of a good mathematical model of the overall system. How can we define a "good" adaptive suspension or a "good" air conditioning system? Reference can only be made to the average of the approval grades in a panel of trained drivers (the average driver). How much would it cost to build a reliable model of a car taking into account, say, all the conditions of the road and of the tires?

This considerations naturally lead to the use of "intelligent" controllers, see e.g. (Aurrand-Lions et al., 1993; Vachtsevanos et al., 1993; Hampo, 1994; Weil et al., 1993; von Altrok and Krause, 1993), which can cut the time-to-prototype and the development cost of new components. Also important, in the case of rule-based controllers, is the easy mantainability of the system. A well structured bank of rule is generally more readable of a mathematical algorithm and upgrades are easier to perform. However, for many of the quoted systems, there is no "expert knowledge" to use. In these cases, a learning approach can be very useful.

In (Hessburg and Tomizuka, 1994), the lateral control of a car along an IVHS is considered. The controller uses three different rule bases. The first implements a feedback law using 125 rules. The inputs are the lateral error, its derivative and the time derivative of the angle between the axes of the car and of the road. The angle itself was not used, being it difficult to measure. The output of this controller is a steering command δ_{fb}.

When determining the steering angle of his car, a common driver acts using a blend of the actual road curvature and of the previewed one, with weights depending on the time to get to the curve. This is included using a second rule base with 52 rules, the output of which is δ_{pr}. Finally, a gain scheduler accepts as input the sum $\delta_{pr} + \delta_{fb}$ and outputs a steering command depending on the car speed.

Sensing is based on magnetic plates embedded in the road, whose magnetization direction encodes the curvature preview. The sampling time of the controller is 0.021 seconds. A thorough experimental validation has been conducted using a real car running at speeds between 30 and 60 km/h. The errors are in the order of 10 cm and are smaller or equal to those obtained with a linear quadratic controller. However no mathematical model has been used to design the fuzzy controller, whilst it is needed for the conventional one.

The sensing problem is addressed in (Pomerlau, 1994). Here a neural network is used to recognize the road edges from a vision system. The estimation error is continuously monitored and is used to issue a warning when the recognition is not reliable.

As a different example of applications, we can consider the injection control for a car engine (Majors et al., 1994). In this case very little knowledge about the control of the system is previously available and a neural approach is undertaken. The design goal is to take constant the air/fuel ratio A/F in a range $\pm 1\%$ to reduce pollution. A CMAC (cerebellar model articulation controller) is used. One of its characteristics is the possibility to make it more or less precise, and more or less memory greedy, by varying a generalization parameter.

The engine state is represented by the intake manifold pressure and by the rotation speed. Its output is the reading provided by the A/F sensor, a device which, being located along the exhaust pipe, measures the air/fuel ratio with some delay.

The training of the network is performed in two steps, the first uses an engine simulator, the other data from real scenarios. As a result the CMAC, implemented on a 386 portable computer, gives the required performances, both at steady state and during fast transients. It also has a very fast convergence rate, being able to learn from a new scenario in just four iterations. In the same conditions, the standard injection system shows much higher errors.

5. CONCLUSIONS

The paper analyzes some representative cases of application of soft computing methodologies to the control of mobile robots and autonomous vehicles. In general, their success is proportional to the complexity of the considered systems and to the uncertainties of its behavior. Indeed, they need no mathematical model of the controlled system and can easily incorporate the qualitative information and the expert knowledge that is often unused in more conventional designs. Thus, controlling a linear or a nonlinear system makes no methodological difference.

Rule-based (fuzzy) controllers show also clear benefits during their operative life. The rule base is almost self-documenting and rules can be easily modified to adapt the controller to new situations or to improve its performances. As a further benefit, they can easily incorporate exception handling situations.

The main advantage in using neural networks or genetic algorithms is their ability to learn from examples. When little previous knowledge on the system (or on the sensor) behavior is available, a learning approach can be very appealing.

However all these are not "magic" tools. They cannot escape from the structural limitations which characterize mobile robots or from the typical problems that affect navigation in unstructured environments, problems that, not surprisingly, are indeed of qualitative nature. Moreover, they generally require a higher computing power than conventional controllers. In this respect, a remarkable gain in terms of efficiency can be found in properly engineered systems.

According to (Thomas and Armstrong-Hélouvry, 1995), we should consider the information equity for the controlled system, i.e. the difference between the value of information and the cost of that information. In many of the considered cases the cost of a crisp information is very high and its utility questionable, as classical methods cannot be easily applied. On the contrary, fuzzy, qualitative information is almost free and the soft-computing methods allow its effective use.

6. REFERENCES

Aguilar, J.M., and J.L. Coutreras-Vidal (1994). Navite: a neural network system for sensory-based robot navigation. In: *Proc. of World Congress on Neural Networks*.

Aurrand-Lions, J., M. de Saint Blancard, and P. Jarri (1993). Autonomous intelligent cruise control with fuzzy logic. In: *Proc. of EUFIT'93*.

Badreddin, E., and M. Mansour (1993). Fuzzy-tuned state-feedback control of a non-holonomic mobile robot. In: *Proceedings of the 12th Triennial World Congress of the International Federation of Automatic Control*. Vol. 3.

Bell, D., J. Borenstein, S. Levine, Y. Koren, and A. Jaros (1994). An assistive navigation system for wheelchairs based upon mobile robot obstacle avoidance. In: *Proc. of 1994 IEEE Conf. on Robotics and Automatio*.

Borenstein, J., and Y. Koren (1991). The vector field histogram-fast obstacle avoidance for mobile robots. *Trans. on Robotics and Automation* **7**(3), 278–288.

Braunstingl, R. (1995). Fuzzy logic wall following of a mobile robot based on the concept of general perception. In: *Proc. of ICAR '95*.

Cox, I. J. and Wilfong, G. T., Eds.) (1990). *Autonomous Robot Vehicles*. Springer-Verlag, Springer-Verlag.

De Luca, A., and G. Oriolo (to appear). *Modelling and control of nonholonomic mechanical systems*. Chap. 7. Springer-Verlag-CISM.

de Wit, C.C., and O.J. Sordalen (1992). Exponential stabilization of mobile robots with nonholo-nomic constraints. *IEEE Trans. on Automatic Control* **37**(11), 1791–1797.

Demel, P., S. E. McCormac, and A. Uhl (1995). A straightforward fuzzy collision avoidance strategy for autonomous vehicles. In: *Proc. of EUFIT'95*.

Deng, Z., and M. Brady (1993). Dynamics and control of a wheeled mobile robot. In: *Proc. of 1993 American Control Conference*. Vol. 2, pp. 1830–34.

Hampo, R. (1994). Ic engine misfiring detection algorithm generation using genetic programming. In: *Proc. of EUFIT'94*.

Hessburg, T., and M. Tomizuka (1994). Fuzzy lateral control for lateral vehicle guidance. *IEEE Contr. System Magazine* **14**(4), 55–63.

Kosko, B. (1992). *Neural networks and Fuzzy Systems*. Prentice-Hall, Prentice-Hall.

Latombe, J. C. (1991). *Robot Motion Planning*. Kluwer Academic Publ., Kluwer Academic Publ.

Liu, K., and F.L. Lewis (1992). Application of robust control techniques to a mobile robot system. *Journal of Robotic Systems* **9**(7), 893–913.

Majors, M., J. Stori, and Dong il Cho (1994). Neural network control of automotive fuel-injection systems. *IEEE Contr. System Magazine* **14**(3), 31–36.

Martinez, J., A. Ollero, and A. Garcia Cerezo (1993). Fuzzy strategies for path tracking of autonomous vehicles. In: *Proc. of EUFIT'93*, p. 24.

Nishimori, K., and et al. (1994). Automatic tuning method of membeship functions in simulation of driving control of a model car. In: *Proc. of 1993 Int. Joint Conf. on Neural Networks*, pp. 2937–2940.

Oriolo, G., G. Ulivi, and M. Vendittelli (1994). Motion planning with uncertainty: Navigation on fuzzy maps. In: *Proc. of IFAC SYROCO 94*.

Pin, F., and Y. Watanabe (1993). Using fuzzy behaviors for the outdoor navigation of a car with low-resolution sensors. In: *Proc. of 1993 IEEE Int. Conf. on Robotics and Automation*, p. 548.

Pin, F., H. Watanabe, J. Symon, and R. Pattay (1992). Autonomous navigation of a mobile robot using custom designed qualitative reasonong vlsi chips and boards. In: *Proc. of 1992 IEEE Int. Conf. on Robotics and Automation*, p. 123.

Poloni, M., G. Ulivi, and M. Vendittelli (1995). Fuzzy logic and autonomous vehicles: experiments in ultrasonic vision. *Fuzzy Sets and Systems* **69**, 15–27.

Pomerlau, D.A. (1994). Reliability estimation for neural network based autonomous driving. *Robotics and Autonomous Systems* **12**(3), 113–119.

Reignier, P. (1994). Fuzzy logic techniques for mobile robot obstacle avoidance. *Robotics and Au-*

tonomous *Systems* **12**(3), 143–153.

Saffiotti, A., E. Ruspini, and K. Konolige (1993). Blending reactivity and goal-directedness in a fuzzy controller. In: *Proc. of 1993 IEEE Conf. on Fuzzy Syst.*, pp. 134–139.

Samson, C. (1993). Time-varying feedback stabilization of car-like wheeled mobile robots. *International Journal of Robotics Research* **12**(1), 55–64.

Santos, V., J.G.M. Goncalves, and F. Vaz (1994). Perception maps for the local navigation of a mobile robot: a neural network approach. In: *Proc. of 1994 IEEE International Conference on Robotics and Automation*, pp. 2193–2198.

Sehad, S., and C. Touzet (1994). Self-organizing map for reinforcement learning: obstacle-avoidance with khepera. In: *Proc. of 1994 From Perception to Action Conference*, pp. 420–423.

Shibata, T., and T. Fukuda (1993). Intelligent motion planning by genetic algorithm with fuzzy critic. In: *Proc. of 1993 IEEE Int. Symp. on Intelligent Control*, pp. 565–570.

Simpson, R., S. Levine, D. Bell, and L. Jaros (1994). The applicability of neural network in an autonomous mode selection system. In: *Proc. of RESNA '94*.

Sugeno, M., and K. Murakami (1985). *An experimental study on fuzzy parking control using a model car.* p. 125. Elsevier Science Publ.

Tanaka, K. (1994). Advanced fuzzy control of a trailer type mobile robot-stability analysis and model-based fuzzy control. In: *Proc. of Sixth International Conference on Tools with Artificial Intelligence*, pp. 205–211.

Thomas, D. E., and B. Armstrong-Hélouvry (1995). Fuzzy logic control — a taxonomy of demonstrated benefits. *IEEE Proceedings* **83**(3), 407–421.

Ulivi, G. (1995). Fuzzy logic and the control of a wheelchair for disabled people. In: *to appear on Proc. of Wilf '95*.

Vachtsevanos, G., S. S. Farinwata, and D. K. Pirovulou (1993). Fuzzy logic control of an automotive engine. *IEEE Contr. System Magazine* **13**(3), 62–86.

von Altrok, C., and B. Krause (1993). Fuzzy logic and neurofuzzy technologies in embedded automotive applications. In: *Proc. of EUFIT'93*.

von Altrok, C., B. Krause, and H. Zimmerman (1992). Advanced fuzzy logic control of a model car in extreme situations. *Fuzzy Sets and Systems* (48), 41–52.

Wang, L. X., and J. M. Mendel (1992). Generating fuzzy rules from numerical data with applications. *IEEE Trans. System, Men, and Cybern.* **SMC-22**, 1414–1427.

Weil, H., G. Probst, and F. Graf (1993). Fuzzy shift logic for an automatic transmission system. In: *Proc. of EUFIT'93*.

Yen, J., and N. Pfluger (1995). A fuzzy logic based extenson to payton and rosenblatt's command fusion method for mobile robot navigation. *IEEE Trans. System, Men, and Cybern.* **25**(6), 971–978.

Beom, H. and Cho, H. (1995). A sensor-based navigation for a mobile robot using fuzzy logic and reinforcement learning, *IEEE Trans. System, Men, and Cybern.* **25**(3), 464-477.

ACKNOWLEDGMENTS

This work was partially supported by Cassa Rurale ed Artigiana di Roma

NAVIGATION WITH REACTIVE BEHAVIOURS FOR THE AURORA MOBILE ROBOT[1]

**A. Mandow*, V. F. Muñoz*, J. Gómez-de-Gabriel*,
J. L. Martínez* and A. Ollero****

* *Departamento de Ingeniería de Sistemas y Automática. Universidad de Málaga.*
Plaza el Ejido s/n, 29013 Málaga (Spain).
Fax: (+34) 5 213-14-13; Tel: (+34) 5 213-14-18; E-mail: tony@tecma1.ctima.uma.es

** *Departamento de Ingeniería de Sistemas y Automática. Universidad de Sevilla.*
Avenida Reina Mercedes s/n, 41012 Sevilla (Spain).
Fax: (+34) 5 455-68-49; Tel: (+34) 5 455-68-71; E-mail: aollero@cartuja.us.es

ABSTRACT. This paper presents the task-oriented navigation system developed for AURORA, a new mobile robot fully designed and built at the University of Málaga. AURORA must perform purposeful tasks in constrained environments with a high obstacle density. The solution is based on the information provided by a set of on-board ultrasonic sensors whose configuration was chosen according to such kind of environments. These sensed data are used by a set of reactive navigation behaviours as well as by the navigation control system, which decides when each behaviour must be activated according to the specified task.

Keywords: Mobile robots, reactive control, task-oriented autonomous navigation

1. INTRODUCTION

This paper describes a solution for mobile robot autonomous navigation, focusing on task-oriented navigation in environments where obstacle density is so high that robotic motion is notably constrained. Such working environments, that will be referred to as constrained or intricate, are common in production facilities where productivity depends on a good use of available space. Such is the case of warehouses, where goods are temporarily stored, leaving just the space needed to take them in or out by pallet transport, a work suitable for mobile robots. Such industrial environments cannot usually be adapted for robotic use without a loss of efficiency and consequently a flexible solution is desirable.

The existing methods to deal with mobile robot navigation can be broadly classified depending on their use of planned or reactive approaches. A planned navigator has to find a trajectory which achieves the task goals by using the information provided by the sensing system and preliminary knowledge about the environment. This methodology needs an updated environment map and a accurate position estimation method. On the other hand, much of the work on reactive navigation of mobile robots has been inspired by the layered control system of the subsumption architecture (Brooks, 1986). Recent trends have focused on the integration of reactive low level behaviours with high level planning in order to obtain speed and reliability in task-oriented navigation (Arkin, 1989; Gat, 1993; Soldo, 1990).

The solution adopted for AURORA consists of a supervisor sequential controller that executes a set of reactive navigation and operation behaviours according to a high level task introduced by the user, which is used by the system to infer a qualitative description of the environment.

The paper is organized as follows. A discussion of the aims and constraints of the navigation system is presented in section 2. A brief description of AURORA and its sensing system is offered in section 3. Section 4 follows with a discussion of the

[1] This work has done within the framework of the Fuzzy Algorithms for Control (FALCON)-working group funded by the ESPRIT Program of the E.C., and has been partially funded by the *P.N.I.C. Almeria-Levante* (*Junta de Andalucia*) and CICYT TAP93-0581.

navigation control system and its implementation. The final sections are devoted to present experimental results and conclusions.

2. TASK-ORIENTED NAVIGATION IN CONSTRAIND ENVIRONMENTS

Navigating in warehouses and other industrial environments requires a considerable reactive capability. On account of productivity, they do not leave much free space for robot navigation. Such a kind of environment can be referred to as constrained or intricate, since the vehicle must navigate through areas where the surrounding objects are so close that they constrain its trajectory. Because of this, the adoption of a classical planned approach (Lozano-Pérez and Wesley, 1979) would pose three main practical disadvantages for constrained autonomous navigation: First, an explicit, up-to-date map description of such dynamic environments would be necessary; secondly, even if a map can be maintained, the robot would need an accurate position estimation method, which demands costly sensing and computing systems; and thirdly, if odometry alone is used for position estimation, the robot may not dare traversing intricate areas on account of the uncertainty accumulated during long trajectories.

On the other hand, purely reactive approaches alone do not suffice, since they are not very adequate to accomplish the objectives of real missions. Consequently, although the basic navigation behaviours can be predominantly reactive, a plan with a high level of abstraction is necessary for task-directed navigation.

Thus, a task is specified as the sequence of navigation behaviours the robot must follow in order to accomplish its mission, i.e. a high level plan. Besides, dynamic industrial environments are fit to be qualitatively described through task specification. In other words, without an explicit representation of the environment, an implicit description of it can be inferred from a sequence of instructions such as "follow the wall on the right; at the end of it turn left and follow the corridor," and so on. During execution, each behaviour can be considered as a state. The robot will change its current state to the next behaviour in the sequence, according to a condition detected through sensed information. Thus, instead of showing a purely reactive behavior, the robot can perform purposeful activities according to a strategic plan (Ollero et al., 1995a).

3. THE AURORA MOBILE ROBOT.

AURORA is an autonomous wheeled mobile robot for greenhouse operations fully designed and built at the Málaga University (see Fig. 1). In particular, the original aim of AURORA is to perform operations in greenhouses, a clear example of constrained environment on account of productivity, which means that it must also cope efficiently with all the harsh navigation conditions presented by agronomic environments. In order to be competitive, AURORA must be capable of operating autonomously in a variety of different greenhouses without imposing any alterations on them at all.

The mechatronic system consists of an octagonal mobile platform that accommodates a spraying device, the power system, standard electronic and computer enclosures, and a variety of sensors for intelligent spraying and navigation. Its dimensions, constrained for the ability to navigate in narrow greenhouse corridors, are 80 cm in width and 140 cm in length.

The locomotion system is a modification of the RAM-1 dual configuration (Ollero et al., 1993) that renders high maneuverability (zero turning radius), which is essential in constrained environments. The main differences with the RAM-1 locomotion system are the increase of the torque, the decrease of top speed from 1.7 m/sec to 0.8 m/sec, all 4 wheels have suspension, and wheel diameter is 40 cms. Besides, in order to obtain more autonomy and disposition to work in agricultural environments, the power system is not based on DC batteries, but on a fueled on-board AC generator. A detailed description of AURORA can be found in (Ollero et al., 1995b).

The implementation of reactive navigation behaviours and the navigation controller relies heavily on local sensor information. Therefore, its efficiency will depend on the type and configuration of the sensing system of the robot. Ultrasonic sensors offer a good and relatively low-cost solution. They can be adjusted to work at different range distances and can thus be combined to sense both the immediate surroundings (a few centimeters) and mid-range distances (a few meters) from the robot. Moreover, ultrasonic sensed

Fig. 1. The AURORA mobile robot.

information is frequently updated and requires little processing, properties that are especially useful for reactive control. Instead of a classical homogeneous sonar ring, AURORA's configuration of ultrasonic sensors has been chosen considering the characteristics of constrained environments. Thus, a combination of three different types of ultrasonic sensors has been placed in the front half of the robot, covering the ranges shown in Fig. 2. The number and types of sonars that have been used are the following: 4 Short-Range Digital (SRD) sensors, 2 Mid-Range Digital (MRD) sensors, and 4 Mid-Range Analog (MRA) sensors with adjustable orientation.

4. NAVIGATION CONTROL

The architecture of the autonomous navigation system developed for AURORA is shown in Fig. 3. This is part of the AURORA control architecture, which is fully described by Gómez de Gabriel *et al.* (1996). Navigation control takes place at the three uppermost levels of this architecture: user, supervisor and reference generation.

4.1. User level

At the top level, a sequential task plan is introduced through the user console. This task specification is meant to be a high level description of the mission to be performed by the robot at run time. Since the purpose of the robot is not navigation in itself, the task must also determine the special operations the robot must complete along its path.

Task specification takes place through a user interface incorporated on the robot that is composed of a two-line display and a 16-key pad. This interface allows the edition of a sequence of behaviours for both navigation and operation. For navigation, a different

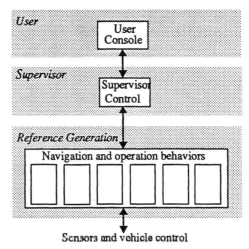

Fig. 3. Autonomous navigation system architecture.

key has been assigned to each possible behaviour, and programming is just as easy as pressing the buttons in the desired sequence, which leads to a program of the type "follow the wall at the left; turn left; follow the corridor...". There is also the possibility of introducing some approximate notions of odometry by using a numerical input (e.g. "follow the wall at least for 2 meters," or "turn left up to 90°").

Regarding operation, AURORA carries an on-board spraying device that must be activated and deactivated depending on the current stage of the navigation task while operating inside greenhouses. This is specified by means of a flag in each instruction of the task (e.g. "Follow the corridor *while operating*"). Similarly, any other operations should also be specified in the sequence. The way these operations are specified will depend on their complexity. In particular, relatively simple operations, like spraying or pallet handling, have been considered. The introduction of coordinated robotic arm operations, for instance, would require further study and, probably, a more sophisticated interface.

4.2. Supervision

This level of the control architecture is in charge of supervising the execution of the task by coordinating the activation and deactivation of the navigation and operation behaviours of the reference generation level. The supervisor keeps track of the current state of the task by managing a flow of events with the reference generation level. It must also take the appropriate actions if any unexpected event takes place during execution.

The actual implementation of navigation control in AURORA is basically of a sequential nature. Since there is a correspondence between the navigation behaviours defined in the reference generation level and the steps that can be programmed in the task, the function of this supervisor is to keep track of the execution of the task by deciding the next behaviour in

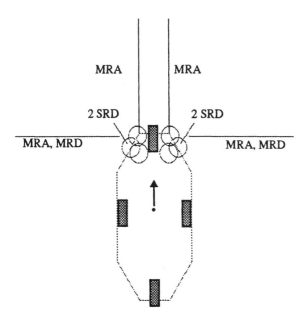

Fig. 2. Operating range limits of ultrasonic sensors.

the sequence to be activated when the current one sends a termination event. A sequential control approach, besides its simplicity, offers a natural and effective way for navigation control. Nevertheless, operation and safety behaviours, are concurrently executed and managed by the supervisor during navigation execution.

The supervisor contains a table with all the possible conditions necessary to change from one behaviour to another. For instance, between a "follow corridor" and a "turn left" behaviour, the repeated detection of a free space by the lateral MRA and MRD sensors is necessary.

4.3. Reference Generation

The objective of this level is to generate the appropriate values to be passed to the lower levels of the control architecture. It is composed of a set of independent agents, called behaviours, each of which is specialized in accomplishing a particular basic action, such as following a corridor or turning round a corner. Each behaviour produces a set of outputs based on a particular interpretation of sensed information from the environment and the robot, provided by the lowest levels of the control architecture.

The reference generation level has been implemented as a set of independent reactive behaviours. The navigation references produced by this level are the desired velocity and curvature of the vehicle. The set of behaviours and their quality will depend on the environment where the robot has to work and the kind of operations it has to perform. Moreover, existing behaviours can be easily modified and new ones can be added without affecting the rest of them. Apart from navigation, this level contains behaviours for operation, safety, etc.

Because of the nature of ultrasonic sensors, which provide numerous erroneous readings, the navigation behaviours must make use of redundancy in order to reliably accomplish their aims. That redundancy is achieved both by repeating readings in each sensor before taking an action, and by integrating the results of different sensors covering the same area.

The main features of AURORA's navigation behaviours, which have been implemented by means of rules, are discussed below.

Corridor Following. This behaviour uses information from sensors at both sides of the robot. Its aim is to maintain the symmetry between them. At each side, the information from the SRD and MRA sonars (3 devices in total) is added to detect if an object is very close to that side of the robot (within a 15 cm range). The robot will be considered centered

is both sides provide the same interpretation, either both clear (a relatively wide corridor) or both close (a constrained corridor, as the example depicted in fig. 4.a). In such cases a zero valued reference is produced. An appropriate control reference of the curvature will be given when a difference is detected between both sides, causing the robot to turn to the clear side. When symmetry is regained a slight compensatory curvature reference is given with the opposite sign in order to avoid the robot to oscillate excessively from one side to the other. It must be noted that while working in constrained environments, there is a limit in narrowness for safe navigation (see Fig. 4.b-c), which is the width of the robot (0.8 m) plus the range of lateral detection (0.15 m), that is 0.95m. In corridors narrower than that, the robot would always be considered centered (objects at both sides), even if it was in contact with one of the walls. That condition can be detected from the analog data of the side MRAs.

Wall following. The wall following behaviour is similar to that for corridor following. It uses only the lateral sensors corresponding to the side of the wall. A control reference of curvature is provided if the wall comes into the short range of the sensors (too close) or when a particular distance is detected through the MRA analog sensor (too far). The intermediate positions produce zero curvature.

Turning. For turning, a combination of local odometry and reactive sensing has been implemented. Odometry is used as a guide to limit the angle to be turned. However, if a close object is detected by the side sensors before that angle is completed, the turning will also be finished. The turning curvature must be specified beforehand; in the case of constrained environments, the default value is that which minimizes the area covered by the robot during the turn, that is a turning radius situated at the inner edge of the robot (in the case of AURORA, 0.4 m).

Safety. Executing concurrently with the navigation behaviours presented above, the safety behaviour contains a set of rules to avoid collision. The front MRA sensors are used to detect any object in the way

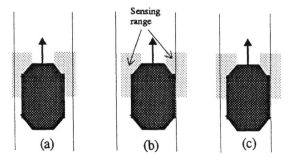

Fig 4. Corridor following. Cases (a) and (b) represent a 'centered' and a 'left' status, respectively. Case (c) is unsafe, since it would be considered as centered.

of the robot during task execution. In its current implementation, AURORA just stops and waits for the object to be removed, but an obstacle avoidance algorithm could be easily integrated here. It must be said that obstacle avoidance has not been considered in constrained environments because of practical reasons: for instance, while operating inside a greenhouse there is usually no space left to avoid the obstacles on the way.

5. EXPERIMENTS

In order to illustrate AURORA´s navigation system a simple example has been set out (see Fig. 5). Fig.6 shows a sequence of images from the actual experiment, in which AURORA navigates autonomously through a recreated storage environment using only the information provided by the ultrasonic sensors. The wall on the right side of the images was a flat surface, but the one on the right was an uneven one, composed using cardboard boxes of different sizes and leaving empty spaces between them, as it is sketched in the diagram of fig. 5.

The program introduced through AURORA´s keypad is composed of the following steps:

1.- Follow corridor.
2.- Turn Left up to 90°.
3.- Follow the wall at the left.

The first two images in Fig. 6 show how the robot follows the corridor. Slight changes of curvature are the reactive response to the detection of an uncentered status by the "follow corridor" behaviour (see Fig. 7). In the last two images, the "turn left" behaviour is activated when a free space is sensed at the left, as seen from the robot, and it approximates and turns to the left, respectively. The moment of the 90° turn is expressed in Fig. 7by the peak of 2.5/m of curvature. The speed and turning curvature for each behaviour are specified beforehand (speed was 0.4 m/s in this particular task).

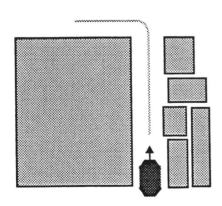

Fig. 5. Experiment layout

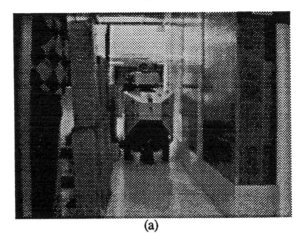

(a)

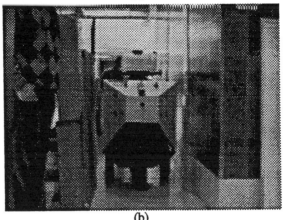

(b)

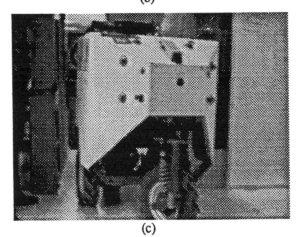

(c)

(d)

Fig. 6. AURORA performing an autonomous mission.

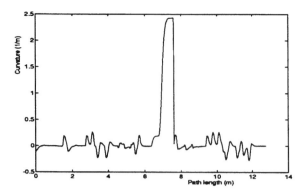

Fig. 7. Evolution of curvature during the experiment.

Fig. 6.a and 6.b give an idea of how constrained the trajectory of the robot can be. Because of this lack of space, oscillations can mean collisions with the walls, and therefore control actions (through curvature) when a deviation is detected must be very smooth. Once the robot is considered to be centered again, a slight compensatory action, in the opposite direction, takes place in order to straighten the trajectory and avoid oscillations from one side to the other.

Furthermore, the uneven shape of the walls is filtered by not using sensed information immediately, but redundantly for short periods of time. In Fig. 6 the sensor array of the robot can be clearly seen.

After several tests with this setup, AURORA repeatedly detected the same environmental landmarks specified in the task (corridor, turning point and wall), accomplishing the mission independently of the robot's starting position or the exact path followed.

6. CONCLUSIONS

The autonomous navigation system presented in this paper is not purely reactive. The reason of this is that real world robots must have some kind of plan "in mind" so that they can perform some particular and useful task. Purely reactive navigation represent the basic survival and navigation behaviours for the robot, but must be completed with task-driven supervision if what is required is something more than a roaming robot that does not bump into things.

This idea has been implemented as a supervisor that sequences the activation of basic navigation behaviours according to a high level sequential task. A sequential control approach, besides its simplicity, offers a natural and effective way for navigation control. Nevertheless, the architecture is flexible enough to support a combined output value from concurrent navigation behaviours in the reference generation level (e.g. behaviours could be overlapped for a period of time instead of changing abruptly), according to a fuzzy activation controller (Ruspini *et al.*, 1995).

Moreover, a first analysis has already been done aiming at the incorporation of fuzzy techniques in the design of navigation behaviours that make full use of the disparate information provided by the ultrasonic sensors. This way, the emergent behaviour of the robot could be enhanced for environments with not so constrained areas (particularly office-like).

7. REFERENCES

Arkin, R. C. (1989) "Motor Schema Based Mobile Robot Navigation". *International Journal of Robotics Research*, Vol 8, pp. 92-112.

Brooks, R. A. (1986), "A Robust Layered Control System For a Mobile Robot". *IEEE Journal on Robotics and Automation.* Vol. RA-2, No. 1, pp. 14-23.

Gat E. (1993). "On The Role of Stored Internal State in the Control of Autonomous Mobile Robots". *AI Magazine*, pp. 64-73, Spring 1993.

Gómez de Gabriel, J., Martínez, J.L., Mandow, A., Muñoz, V.F., Ollero, A., (1996) "The AURORA Mobile Robot Architecture", submitted to IFAC World Congress. San Francisco.

Lozano-Pérez T. and Wesley M. A. (1979). "An Algorithm for Planning Collision-Free Paths Among Polyhedral Obstacles. Mobile Robot Trajectory Planning With Dynamic and Kinematic Constraints". *Communications of the ACM*, Vol. 22, No 10.

Ollero A., Simón A., García F. and Torres V. (1993). "Mechanical Configuration and Kinematic Design of a New Autonomous Mobile Robot". *Intelligent Components and Instruments for Control Applications*, pp. 461-466. Pergamon Press.

Ollero A., Mandow A., Gómez-de-Gabriel J. and Muñoz V.F. (1995a). "Autonomous Mobile Robot Operation and Navigation in Industrial Environments". *Robotics and Computer Integrated Manufacturing*. Vol. 12. Kluwer.

Ollero, A., Martínez, J.L., Simon, A., Gómez-de-Gabriel, J., Muñoz, V.F., Mandow, A., García-Cerezo, A., Garcia, F., and Martínez, M.A., (1995b). "The Autonomous Robot for Spraying AURORA," IARP 95, Robotics in Agriculture and the Food Industry, Toulouse, France.

Ruspini, E.H., A. Saffioti, K. Konolige. (1995) "Progress in Research on Autonomous Vehicle Motion Planning." *Industrial Applications of Fuzzy Logic and Intelligent Systems*. Edited by J. Yen, R. Langari and L. A. Zadeh, IEEE Press.

Soldo, M. H. (1990). "Reactive and Preplanned Control in a Mobile Robot". IEEE Int. Conf. on Robotics and Automation, pp. 1128-1132.

FUZZY LOGIC-BASED NAVIGATION FOR AN AUTONOMOUS ROBOT

Pierre Yves Glorennec*

INSA de Rennes, 35043 RENNES CEDEX, FRANCE. E-mail: glorenne@irisa.fr

Abstract. This paper describes a fuzzy logic-based real-time navigation system for a mobile robot in an unknown environment. The fuzzy rule base is constructed from a few typical situations for which we can deduce appropriate control actions easily. These rules are tuned using a reinforcement learning scheme based on Fuzzy Q-Learning.

Key Words. Fuzzy Logic, obstacle avoidance, edge-following, goal-seeking, reinforcement learning.

1. INTRODUCTION

Navigation of an autonomous mobile robot in a dynamically changing unknown environment has recently received attention. This task combines obstacle avoidance, edge-following and goal-seeking behaviors. If the environment is known perfectly a path can be planned off-line for the robot to follow (Martinez, 1994). But in reality, this environment may change: moving objects, unexpected obstacles, loss of accuracy in the computer representation of the environment. Reactive control systems have been designed as alternative methods for a dynamic environment. The usual methods for implementing reactive control include potential fields (Khatib, 1986; Brooks, 1989), neural networks (Kozake, 1991; Fagg, 1994) or evolutionary stategies (Bonarini, 1994; Lin, 1994).

These methods have the following disadvantages:

- for potential fields: difficulty in finding the right parameters and thresholds, unstable oscillations in some cases, local minima,
- for neural networks: difficulty in obtaining training data, length of the training phase,
- for evolutionary strategy: computational complexity and length of the training phase.

In all these methods, it is difficult to introduce some prior knowledge which however is easily deduced from human experience. This experience can be summarized with a few meta rules such that, for example:

- if there is an obstacle, execute a maneuver to go around it,
- if the way is clear, drive to the goal by steering the wheels appropriately.

These rules are easily embedded into Fuzzy Inference Systems (FIS) and many researchers proposed fuzzy logic-based navigation (Beom, 1995; Bonarini, 1994; Li, 1994; Liu, 1994; Reignier, 1994). In (Beom, 1995), obstacle avoidance and goal-seeking are two different tasks and a switching mechanism is used to choose the appropriate behavior, with the risk of oscillations. The authors use a reinforcement learning algorithm to construct the rule base. In (Bonarini, 1994; Li, 1994; Liu, 1994; Reignier, 1994), the fuzzy controller combines obstacle avoidance and goal-seeking tasks and executes the fusion of different basic behaviors.

In this paper, we propose a fuzzy controller, initialized using available behavioral knowledge. A reinforcement learning algorithm, using the Fuzzy Q-Learning paradigm (Glorennec, 1994), allows online fine tuning of the controller parameters.

2. THE ROBOT

We are working on a low-cost robot with ultrasonic sensors that give low resolution and low precision. Therefore we need a preprocessing stage in order to increase the accuracy of sensor data. As a result of this preprocessing, we obtain the distance to the obstacles in three directions, denoted as d_{left}, d_{front} and d_{right}. The linguistic variables $near\,(NR)$ and $far\,(FR)$ are chosen to fuzzify these three distances, see figure 1.

In each direction, we define two positive parameters, d_m and d_s:

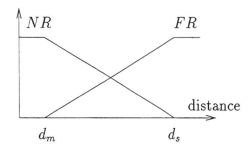

Fig. 1. Fuzzification of the distances

d_m = minimum distance to an obstacle,
d_s = security distance

If the distance is greater than d_s, we consider that the robot can move towards the goal freely.

Let $x \rightarrow \mu_A(x)$ be the membership function of the fuzzy set A and $x \rightarrow S_{a,b}(x)$ the "saturation" function defined by:

$$S_{a,b}(x) = \begin{cases} 0 & \text{if } x < a \\ \frac{x-a}{b-a} & \text{if } a < x < b \\ 1 & \text{if } x \geq b \end{cases} \quad (1)$$

We have:

$$\mu_{FR}(x) = S_{d_m,d_s}(x) \quad (2)$$
$$\mu_{NR}(x) = 1 - \mu_{FR}(x) \quad (3)$$

The robot is moved by controlling the heading velocity, v, and the incremental steering angle, $\triangle\phi$. From the choice of fuzzy input variables, we have to define the behavior of the robot in eight basic situations[1]. For simplicity, we note "(A, B, C)" the statement "d_{left} is A and d_{front} is B and d_{right} is C". Therefore, the eight basic situations are:

$$
\begin{array}{lll}
S_1 & : & (NR, NR, NR) \\
S_2 & : & (NR, NR, FR) \\
S_3 & : & (NR, FR, NR) \\
S_4 & : & (NR, FR, FR) \\
S_5 & : & (FR, NR, NR) \\
S_6 & : & (FR, NR, FR) \\
S_7 & : & (FR, FR, NR) \\
S_8 & : & (FR, FR, FR)
\end{array}
$$

S_1 corresponds to a "cul-de-sac", S_2 to a left corner, S_3 to a corridor etc. Now, starting from human knowledge, we have to define the behavior of

---

[1] In (Reignier,1994) we have also the same basic situations

the robot, using fuzzy rules on the form:

$$\text{if } S_i \text{ then } v \text{ is } v_i \text{ and } \triangle\phi \text{ is } \phi_i \quad (4)$$

with $i = 1$ to 8.

3. THE FUZZY SYSTEM

We have chosen a Takagi-Sugeno FIS with three inputs, d_{left}, d_{front} and d_{right}, and two outputs, v and $\triangle\phi$. The values v_i and ϕ_i in the consequent part of the rules (see equation (4)) are either a fuzzy label reduced to a singleton (e.g. Zero (ZR), Positive Small (PS),...) or a procedure. With strong fuzzy partition on input domains (see figure 1) and product conjunction, the two outputs of the FIS are given by:

$$v = \sum_{i=1}^{8} \alpha_i v_i \qquad \triangle\phi = \sum_{i=1}^{8} \alpha_i \phi_i \quad (5)$$

where α_i is the truth value of rule i given an input vector. If A, B and C are the three fuzzy labels in the premisse part of rule i, we have:

$$\alpha_i = \mu_A(d_{left}) \times \mu_B(d_{front}) \times \mu_C(d_{right}) \quad (6)$$

3.1. Rules for velocity control

The robot is designed to reach the goal with zero speed. Let d_g be the distance to the goal and v_{max} the maximum velocity of the robot. If the way is free, the velocity is given by:

$$v = f(d_g) = v_{max} \times S_{d_0,d_1}(d_g) \quad (7)$$

where the parameters d_0 and d_1 should be determined by learning: if d_1 is big, then the robot is "sluggish" whereas if d_1 is small, then it is "nervous". Let $C = \min(\mu_{FR}(d_{right}), \mu_{FR}(d_{left}))$ be a multiplicative factor used to reduce speed in a corridor. The rules for velocity control are issued from two meta rules:

if d_{front} is NR then **stop**
if d_{front} is FR then **go on**

These meta rules are expanded to the following rule base:

$$
\begin{array}{llll}
\text{if} & S_1 & \text{then} & v \text{ is } ZR \\
\text{if} & S_2 & \text{then} & v \text{ is } ZR \\
\text{if} & S_3 & \text{then} & v \text{ is } C \times f(d_g) \\
\text{if} & S_4 & \text{then} & v \text{ is } f(d_g) \\
\text{if} & S_5 & \text{then} & v \text{ is } ZR \\
\text{if} & S_6 & \text{then} & v \text{ is } ZR \\
\text{if} & S_7 & \text{then} & v \text{ is } f(d_g) \\
\text{if} & S_8 & \text{then} & v \text{ is } f(d_g) \\
\end{array}
$$

In most cases, the actual velocity results from the fusion of several rules by the inference mechanism ans is continuously varying between 0 and $f(d_g)$, decreasing or increasing according to obstacles and goal locations.

3.2. Rules for steering control

The rules for steering control are designed to fusion three different behaviors: obstacle avoidance, edge-following and goal-seeking behaviors. For a given situation, provided by the sensors, the conclusions can differ according to extra considerations. For example, in situations S_1 ("cul-de-sac"), the robot must choose between left or right. The actual choice can be done considering either the goal location or the space needed to turn.

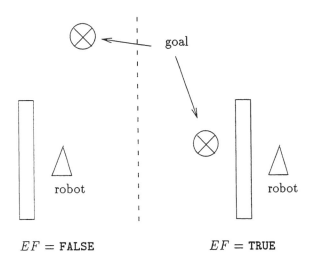

Fig. 2. Following an edge

A boolean, EF, for Edge-Following, is introduced in situations S_4, S_7 and S_8, see Figure 2. If EF is TRUE, then the robot must follow the edge whereas if EF is FALSE it must go away. We define:

$$
\phi_4 = g_1(EF) = \left\{ \begin{array}{ll} PS & \text{if } EF \text{ is TRUE} \\ NS & \text{otherwise} \end{array} \right. \quad (8)
$$

$$
\phi_7 = g_2(EF) = \left\{ \begin{array}{ll} NS & \text{if } EF \text{ is TRUE} \\ PS & \text{otherwise} \end{array} \right. \quad (9)
$$

Table 1 Situation-behavior relationship for S_6.

situation	behavior
$d_{front} = Near$	stop (to avoid the crash)
$d_{left} = d_{right} = Far$	turn right or left

$$
\phi_8 = g_3(EF) = \left\{ \begin{array}{ll} 0 & \text{if } EF \text{ is TRUE} \\ \theta & \text{otherwise} \end{array} \right. \quad (10)
$$

where θ is the angle between the heading direction of the robot and the goal direction.

Therefore, the rules for the steering angle are:

$$
\begin{array}{llll}
\text{if} & S_1 & \text{then} & \Delta\phi \text{ is } \pm Big \\
\text{if} & S_2 & \text{then} & \Delta\phi \text{ is } NB \\
\text{if} & S_3 & \text{then} & \Delta\phi \text{ is } ZR \\
\text{if} & S_4 & \text{then} & \Delta\phi \text{ is } g_1(EF) \\
\text{if} & S_5 & \text{then} & \Delta\phi \text{ is } PB \\
\text{if} & S_6 & \text{then} & \Delta\phi \text{ is } \pm Big \\
\text{if} & S_7 & \text{then} & \Delta\phi \text{ is } g_2(EF) \\
\text{if} & S_8 & \text{then} & \Delta\phi \text{ is } g_3(EF) \\
\end{array}
$$

For example, for rule 6, we have the situation depicted in table 1:

The inference process is the following:

1. Observe the environment: d_{left}, d_{front}, d_{right} and location of the goal.
2. Compute v_i and ϕ_i, $i = 1$ to 8.
3. Compute the truth values, α_i, $i = 1$ to 8.
4. Compute the outputs v and $\Delta\phi$ using equations (5).

4. SIMULATIONS

To illustrate the effectiveness of the proposed method, we have simulated the robot behavior in unknown envireonments. The "room" is a square of 16 × 16 units and the maximum range of the sonars is only 2 units: the robot has therefore a limited sight of its environment.

The Figure 3 shows a robot start position inside a U-shaped obstacle and a goal position located outside.

Without the boolean EF, the robot would be trapped inside this U-shaped obstacle due to local minimum, but as EF is computed TRUE it follows the edges and reaches the goal easily. However we can see some non smooth behavior in the steering angle due to empirical values of fuzzy labels (NS, PS, NB, PB). In the next section, a reinforcement learning scheme is used for fine tuning of these labels.

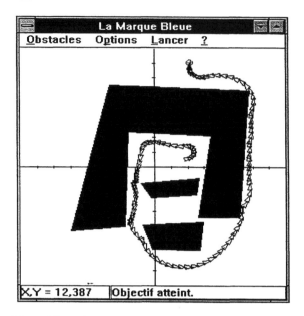

Fig. 3. The robot can escape from the U-shaped obstacle.

5. REINFORCEMENT LEARNING

From human knowledge, we can easily write a first set of rules with approximate conclusions. This initial rule base can be improved on-line, using the reinforcement signal r defined by:

$$r = \begin{cases} -1 & \text{if collision} \\ -0.5 & \text{if distance} < d_m \\ 0 & \text{otherwise} \end{cases} \quad (11)$$

Two main types of method can be used for reinforcement learning: they are based on Adaptive Heuristic Critic (AHC) (Sutton, 1990) or on Q-Learning (Watkins, 1989). In AHC, two online procedures are required: the first procedure is based on the Temporal Difference method (TD) (Sutton, 1989) and estimates an evaluation function; the second one is used to improve the actions. In Q-Learning, a unique procedure estimates the utility of admissible state-action pairs, so that an optimal policy can be determined.

In standard one-step Q-Learning, an agent receives sensory input x from the environment and determines a control action a. A Q-function, $Q(x, a)$, is associated with each state-action pair (x, a): $Q(x, a)$ is "the expected discounted sum of future rewards for performing action a in state x and performing optimally thereafter" (Sutton, 1992).

We have proposed an extension of Q-Learning to FIS (Glorennec, 1994). In our implementation, several agents , which are different outputs of a FIS, are competing to control a process. To each agent, j, is associated a Q-function, $x \rightarrow Q[j](x)$, which is updated as new information is gathered. We have proposed to represent the current estimation of $Q[j]$ as an extra output of the FIS. Therefore, to each agent, j, and to each rule, i, we associate a real-valued parameter, $q[i, j]$, that can be thought of as the quality value of the ith rule used by the jth agent.

More precisely, in our robotic problem, for each agent we define a Q-function for the steering angle.

$$Q[j](x) = \sum_{i=1}^{8} \alpha_i(x) \times q_\phi[i, j] \quad (12)$$

Starting from the rule base defined in section 4, we have built five agents in our simulations, by slightly changing the parameters used in the rules: each agent is a potential fuzzy controller for the robot. These agents are competing to control the robot. As learning proceeds, given a current input vector x, the active agent, A_j, $j = 1$ to 5, is selected with a Boltzmann probability:

$$Proba(A_j | x) = \frac{\exp \frac{Q[j](x)}{T}}{\sum_k \exp \frac{Q[k](x)}{T}} \quad (13)$$

where T is a parameter controlling exploration.

The active agent, j:

1. computes the control values, v and $\Delta \phi$,
2. receives the reinforcement vector, r, from the environment,
3. computes a new estimate of $Q[j]$,
4. updates the parameters $q[i, j]$, $i = 1$ to 8.

In this way, as learning progresses, we obtain both the evaluation of the agents and of the rules they are using. After learning, the optimal policy is to take the agent in each state with the highest Q-function. This is the so-called *greedy policy*.

The figure 4. gives the structure of the used FIS.

6. CONCLUSION

The fuzzy rule base of section 4 gives common sense behavioral rules, allowing the robot to evolve in unknown environments. The rule base has eight rules only but, before using it, we have to set some parameters precisely, e.g.: d_1 in equation (7), modal values of PS, PB,... For fine tuning of these

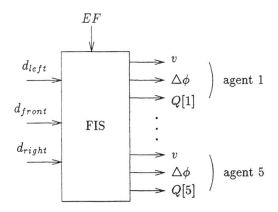

Fig. 4. Structure of the FIS.

parameters, we propose an on-line reinforcement learning method inspired by Q-Learning. In our simulations, we have used five agents with different parameter values.

Based on these simulation results, we are working on a real robot to test the effectiveness of our method in a realistic environment.

7. REFERENCES

Barto A., Sutton R., Anderson C. (1983). "Neuronlike element can solve difficult learning control problems", *IEEE Trans. on SMC*, Vol. 13, Sept. 83.

Beom H., Cho H. (1994). "A sensor-based navigation for a mobile robot using fuzzy logic and reinforcement learning", *IEEE Trans. on SMC*, Vol. 25, March 95.

Bonarini A. (1994). "Learning behaviors represented as Fuzzy Logic Controllers", *Proc. of EU-FIT'94*, Aachen, sept. 94.

Brooks R.A., Flynn A.M. (1989). "Robot being", *Proc. of IEEE/RSJ Int. Workshop on Intelligent Robots and Systems*, Tsukuba, 1989.

Fagg A., Lotspeich D., Bekey G. (1994) "A reinforcement-learning approach to reactive control policy design for autonomous robots", *Proc. of Computational Intelligence Conf.*, Orlando, June 94.

Gardner D., Ashenayi K., Timmerman M., Shenoi S. (1994). "Autonomous Control Hardware for Real-Time Applications", Proc. of FUZZ-IEEE Conference, Orlando, June 94.

Glorennec PY. (1994). "Fuzzy Q-Learning and Dynamical Fuzzy Q-Learning", *Proc. of FUZZ-IEEE'94*, Orlando, juin 1994.

Khatib O. (1986). "Real-time obstacle avoidance for manipulators and mobile robots", *Int. J. of Robotics Research*, Vol. 5, n. 1, 1986.

Kozakiewicz C., Ejiri M. (1991) "Neural network approach to path planning for two dimensional robots", *Proc. of Int. Workshop on Intelligent Robots and Systems*, Osaka, 1991.

Li W., Feng X. (1994). "Behavior fusion for robot navigation in uncertain environment using fuzzy logic", *Proc. of SMC'94*, San Antonio, 1994.

Lin H.-S., Xiao J., Michalewicz Z. (1994). "Evolutionary navigation for a mobile robot", Proc. of EC Conference, Orlando, June 94.

Liu K., Lewis F. (1994). "Fuzzy logic-based navigation controller for an autonomous mobile robot", *Proc. of SMC'94*, San Antonio, 1994.

Martinez-Alfaro H., Flugrad D. (1994) "Collision-free path planning for mobile robots and/or AGVs using simulated annealing", *Proc. of SMC'94*, San Antonio, 1994.

Reignier P. (1994). "Pilotage réactif d'un robot mobile", Thèse à l'INPG de Grenoble, décembre 1994.

Sutton R.S. (1989). "Learning to predict by the method of temporal diferences", *Machine Learning*, 3. pp. 9-44, 1989.

Sutton R.S. (1990). "Integrated architectures for learning, planning and reacting based on approximating dynamic programming", *Proc of Int. Conf. on Machine Learning*, San Mateo, 1990.

Sutton R.S., Barto A., Williams R. (1991). "Reinforcement learning is direct adaptive optimal control"", *Proc of ACC*, Boston, june 1991.

Watkins C. (1989). "Learning from delayed rewards", PhD Thesis, University of Cambridge, England.

A DYNAMIC PATH PLANNER FOR
AUTONOMOUS VEHICLES

J. C. ALVAREZ*, H. LOPEZ* , J. A. SIRGO* and L. SANCHEZ**

*University of Oviedo, Department of Electrical, Electronic, Computers and Systems Engineering, E.T.S. Ingenieros
Industriales e Informáticos, Gijón, SPAIN

**University of Oviedo, Languages and Systems Area, E.T.S. Ingenieros Industriales e Informáticos, Gijón, SPAIN

Abstract. This paper proposes a method for the design of a dynamic path planner which utilizes
fuzzy logic for navigation of mobile robots in uncertain environments. The method consists in
identifying a dynamical system, with a fuzzy logic controller structure, from the examples generated
by an "expert". The generator of examples will be a search algorithm on a known map of the
working environment, that is, a model-based path planning. The dynamic path planner maps the
ultrasonic sensors readings and the odometer measures to the robot control commands –its heading
velocity and steering angle–. The correct mapping is found by identifying a set of fuzzy rules from
desired input-output data pairs. It is wanted with this idea to use the optimality qualities that the
geometrical path planners have, integrating them in an sensor-based architecture, better adapted
for the real-time specifications. The effectiveness of the proposed method is verified by a series of
simulations.

Key Words. Robotic Navigation, Mobile Robots, Fuzzy Control, Path Planning

1. INTRODUCTION

Path planning (PP) consists in the generation of
routes free of colision for mobile robots. This
is one of the most studied problems in robotics,
being able to distinguish different approaches according to the departure conditions:

- for vehicles or for manipulators (2 or 3 dimensions)
- in known environments (given a map) or not
- in static or dynamic environments (with mobile obstacles)

The first studies, for known and static environments, permitted to establish a methodology, ap-
plicable to 2 as well as to 3 dimensions. This
method is based on reformulating the path planning as a geometrical problem: to move a rigid
solid through a n-dimensional euclidian space with
obstacles. And it can still be simplified to the
problem of moving a point in an adequately selected space, the so called space of the Configurations or C space. In practice, the movement of the
robot, is usually limited by cinematic restrictions,
which can reduce the possible roads, by limiting
the accessible C space. The best route is selected
with search algorithms in the states space mentioned above.
These methods, also known as *static*, *open-loop* or
Model-based navigation approaches, continue being intensely studied (Latombe, 1991). They have

the virtue of permitting to determine, in static
and known environments, if the goal is reach-
able and, in that case, which is the best route to
achieve it –according to a previously defined criterion by the designer –. However they present the
drawback of demanding a great quantity of calculation (search algorithms), what makes them im-
proper for the requirements of the real-time navigation (Dai Feng, 1990).
But the real environments are normally not structured, that is to say, unknown and dynamic. Because of this, in practice, the static path planning
has to be combined with other algorithms, either
to avoid unexpected obstacles (for unknown environments), or to detect and predict the situation
of mobile obstacles (for dynamic environments).
These *hybrid approaches* to mobile robot navigation lead to nested and heterogeneous architectures, with modules of different nature and with
different tasks, making the results over different
robots be of difficult generalization (Gat, 1993).
The *Sensor-based* or *dynamic* navigation methods
generate control commands based on sensor data.
They operate without information of a previous
map of the environment. Although dynamic PP
can easily react to obstacles detected in real-time,
they may not be able to reach the goal, even if a
path exists.
The first projects in dynamic PP (Lumelsky and
Stepanov, 1986) were based on algorithms. How-

51

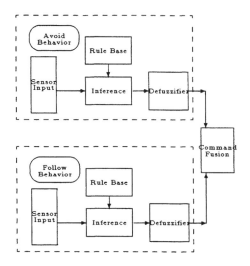

Fig. 1. Behavioral architecture for sensor-based path planning

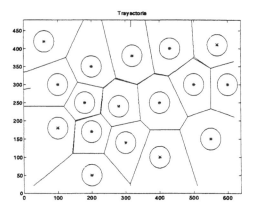

Fig. 2. Simplified Voronoi diagram of a cluttered environment. The central points in the obstacles are the selected significant points, and the lines between obstacles are the roadmap. The model-based path with $q_0 = (143, 110)$ and $q_f = (550, 345)$ is drawn

ever, it exists a clear trend to substitute, wherever it is possible, algorithms by dynamic system structures (Salichs *et al.*, 1993), for two reasons:

1. The homogeneity of the different levels of the robot (planification and control) makes easier the jointly analysis of the whole solution.
2. The Control Theory permits to seek solutions based on non heuristic criteria (such as optimization)

A widely used architecture for dynamic PP is the *behavioral architecture*, which consists of multiple reactive systems. Each one reacts according to the sensor input and contributes to generate a command control (Brooks, 1986). Examples of typical behaviors used in navegation include goal-seeking, wall-following and obstacle-avoidance.

2. FUZZY LOGIC BASED NAVIGATION

In the referred context, the dynamic PP with Fuzzy Logic Controller structure (FLCs) seems a promising alternative (Wang, 1994). The reasons are derived from the general properties of the FLCs:

- Permit to represent and to process the human knowledge –of symbolic character– in a numerical framework.
- They are universal approximators by interpolation among rules
- Present certain hardiness against the uncertainty
- Permits to specify control in local space regions

Though the fuzzy logic methods are –from their theoretical foundations– a question of controversy, their usefulness in the solution of practical engineering problems seems indisputable; for example, it is being used in the inferior levels of the robots control (Ollero and García-Cerezo, 1993). Different dynamic planners, based on behaviors with FLC structure, have been proved (Saffioti *et al.*, 1993; Yen and Pfluger, 1995; Lee and Wang, 1994; Bonarini, 1995). In them, the interest has been focused on defining, from a division of the problem in elemental behaviors – usually goal-seeking and obstacle-avoidance –, the method to generate complex behaviors with a suitable scheme for fusing the control commands (see Fig.1); in this case the primitive behaviours have structure of FLCs, heuristically configured. However, in cases of operating the mobile robot in complex environments, it is difficult to construct rule bases consistently , because there are many situations to be considered.

3. A DYNAMIC PATH PLANNER DESIGN

In this work, a method for the automatic design of a sensor-based navigator based on FLCs models is presented. The aim is to make use of the optimality property that the static path planners have, incorporating it into a purely reactive architecture.

The proposed design method consists of three phases:

1. Implementation of a model-based planner, that optimizes a given criterion for a certain task.
2. Sensorial simulation: obtaining the sensor signals during the trip through the optimal path.
3. Identification of the equivalent dynamic PP, for that task, from the obtained data in the previous simulation.

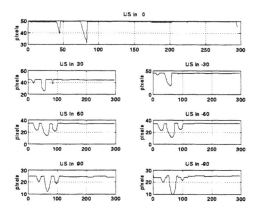

Fig. 3. Sonar readings in the path-following experiment of Fig.2

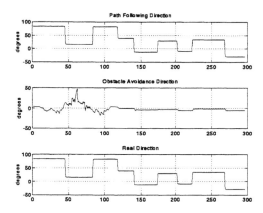

Fig. 4. Path Following experiment, based on the model-based path planner, with $q_0 = (143, 110)$ and $q_f = (550, 345)$. Crisp sensor integration produced by each behavior.

3.1. Model-based path planner

A robot *configuration* is a point of R^2, $q = [x, y]$, that defines its spatial position. The static path planner generates a sequence of collision-free configurations $\{q_{opt}\}$, from the initial and final configurations (q_0 and q_f respectively) of the robot, based on a map of the environment, and optimal in some sense. This planner will be the examples generator for the final identified sensor-based planner.

For this paper a simplified roadmap method was used (Latombe, 1991). The environment map is reduced to its most significant points –usually the center points of the obstacles–, and then the robot free space $\mathcal{C}_{free}$ is retracted on its Voronoi diagram $Vor(\mathcal{C}_{free})$, which acts as the roadmap. The retraction ρ is an application of $\mathcal{C}_{free}$ onto the roadmap $Vor(\mathcal{C}_{free})$.

In Fig.2 an environment, the considered significant points and the voronoi diagram are shown. The paths produced this way are the most distant from the significant points. In this sense it is used the term "optimal path".

The planner proceeds as follows:

1. Select the significant points.
2. Compute the Voronoi diagram $Vor(\mathcal{C}_{free})$.
3. Compute the points $\rho(q_0)$ and $\rho(q_f)$ and identify the arcs of $Vor(\mathcal{C}_{free})$ containing these two points.
4. Search $Vor(\mathcal{C}_{free})$ for a sequence of arcs conected and containing $\rho(q_0)$ and $\rho(q_f)$. The graph is searched using an un-informed A^* algorithm (Dijkstra's algorithm).

3.2. Sensorial Simulation

In this work, the simulation does not intend to be a way of validation or testing of the algorithm, but a part of the method itself. The sensorial simulation consists in obtaining a group of sequences $\{\bar{s}_s\}$ –those which the proximity sensors over the robot would give travelling the route $\{q_{opt}\}$– as well as the measures of position $\{q_s\}$ from the wheels encoders. Both group of measures will constitute the input space to the dynamic navigator. For this phase we have developed a simulator of an holonomic vehicle that moves in a static environment. The robot is equipped with range sensors (a ring of sonar sensors) and encoders in the wheels, with which its position is estimated. Both measures are subject to uncertainties, that have to be addressed in order to obtain a realistic simulation.

Position Estimation. The encoders' signals permit to deduce the robot position in each instant. However these signals contain errors, due to the wheels' slipping as well as quantization errors. Both errors increase while the robot is moving. The problem can be solved in several ways (Talluri and Aggarwal, 1992) : 1) with landmark-based methods; 2) integrating the path (dead reackoning); 3) with standard reference patterns; 4) from a priori information of the environment that has to be recognized with sensing.

The odometry measures are mixed with the goal position q_f in order to select, as navigation inputs, the heading angle ψ and the distance to the goal $z = |q_s - q_f|$. A fuzzy operator converts the crisp input data into the linguistic values, considered as labels of fuzzy sets. This operation implies a robustness, if the hyphothesis that the odometry error is smaller than the ambiguity inherent to the correponding fuzzy value is admitted.

Sonar range sensors. The sonar range sensors play an important role in many applications of robotics. The distance measures that are obtained from them, are affected by echoes, by reflections and the relative orientation between the beam and the obstacle (Borenstein and Koren, 1988). A realistic simulation of these devices has to take into

53

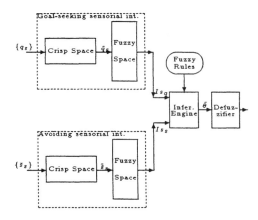

Fig. 5. The structure of the proposed dynamic navigator

account these factors.

In principle we have considered the robot equipped with 7 sensors of high directionality (with an ideal emission cone, $\alpha = 0$). A broadly used model of the uncertainty in this measure, it is to suppose it contaminated with noise characterized by a uniform or a gaussian probability distribution.

Normally the signals of the range sensors are too numerous or complex. To process them in order to deduce behavior rules, can be a highly cost task. In Fig.3 the measures proceding from the seven sonar sensors are shown, corresponding to the model-based path planner ploted in Fig.2. With broad cone sonars, this number can be reduced, but it increases the uncertainty associated with each signal. Because of this, it is necessary to deduce a previous sensorial integration, that reduces the effort of the dynamic navigator.

One possibility is based on a vectorial sum of the repulsive forces, using the sensors readings to compute them. So the planner recieves a vector that tends to move the robot in the direction to distance from the obstacles. This is the so called "potencial field" method. Another one consists on making a weighted sum of de distances reported. Supposing the sonars simetrically disposed around the robot, this constitutes a simplified potencial field method. The third tested possibility is converting each measure into a fuzzy set, and integrating all of them in one with a fuzzy operator, like the max operator.

In Fig.4 the result of the sensor integration for both behaviors is shown. Because it is referred to the experiment explained in the previous figures (model-based navigation), the obstacle-avoid behavior does not affect to the final steering decision θ.

In Fig.5 the proposed navigator architecture is shown. The number of fuzzy rules banks is reduced to one, while the modules of sensorial in-

tegration preserve the behavioral character of the planner architecture. The sensor integration consists of two parts: first, the crisp sensors input space is converted into a fuzzy sensor space; second, the dimension of the input space is reduced to feed the decision making engine. In the two potential-like methods, this reduction is made before the fuzzifing process –reducing the crisp sensor space–, while in the max operator method this is made afterwards –in the fuzzy sensor space–.

Table 1 Numerical results of the related experiments.

	Exp1	Exp2	Exp3	Exp4	Exp5
Coef.					
LeM	295	317	340	311	349
SmM	-53	-54	-26	10	16.9
SM1	35.2	34.1	35.4	35.6	35.3
SM2	17.5	11.2	17.4	22.9	20.1
Min	10.2	9.5	10.2	17.0	15.0

3.3. Simulation Results

In Table 1, are resumed the results of different simulation experiments. A group of quality index of the generated paths was adopted:

1. Lengthness measure (LeM): related to the number of decisions made by the planner to reach the goal.
2. Smoothness measure (SmM): average steering angle.
3. Safety measure 1 ($SM1$): average distance to the obstacles.
4. Safety measure 2 ($SM2$): average minimal distance to the obstacles.
5. Mininal distance (Min): minimal distance to the obstacles reported during the whole travel.

The first experiment consists in the model-based travel reported in the previous figures. So, the robot was sent right ahead through the optimal path (over the Voronoi diagram). The second one uses the simple weighted sum of the sonar readings to make the same mission.

The experiments 3 to 5 compare diferent navigators for the task reflected in Fig.6. Experiment 3 is the model-based path planner, and produces an apriori collision free navigation. Experiment 4 uses the proposed navigator, with a previous integration of the crisp sonar readings -weighted sum- and a linear-like fuzzy rules bank. The path is smoother and more secure than the first one. The fifth experiment uses the fuzzy rules bank identified from the model-based planner, as explained in the next section.

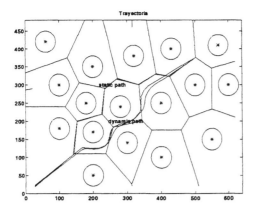

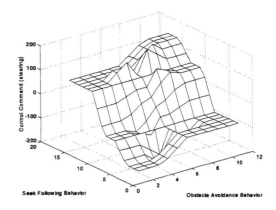

Fig. 6. Graphic results in experiments no. 3 and 4, for paths with $q_0 = (26, 20)$ and $q_f = (502, 376)$.

Fig. 7. An identified fuzzy rule bank

4. FUZZY RULES IDENTIFICATION

The control actions that the dynamic PP generates are the configurations of the robot, in the form of sequence $\{q_r\}$; that is to say, the following position of the path to follow. The sensors information consists of odometer measures $\{q_s\}$ and of the navigation (range) sensors $\{\bar{s}_s\}$. The values of $\{q_r\}$ are related to the response $\{q_{opt}\}$ of the "expert" (that provided us an static PP, as it was already said).

The outlined problem is reduced to an identification of the model underlying to the input ($\{q_s\}$ y $\{\bar{s}_s\}$) and output ($\{q_{opt}\}$) sequences. Those signals result from the sensorial simulation, therefore they are submitted to the mentioned uncertainties.

The model to identify has to be a dynamic system able of handling the different uncertainties that affect the problem:

- The identifition is made with signals coming from a simulation on a deterministic map. It is natural to suppose that there will be deviations between this and the real environment.

- The repetition of experiments can provide different controls to the same states. Given some types of sensors, disposed in a certain way, the quantity of information that collect from the environment will always be less than the one provided by the perfect initial map. The bigger the loss of information is, the further the identified dynamic PP will be from the original "teacher".

The chosen identification method is based on a previous division of the states space X (from sensing) and U (from outputs) in fuzzy partitions, and to generate a rules bank in the form of table (Wang and Mendel, 1992).

In Fig.7 it is collected, in its control surface, the identified fuzzy map (for the steering control command θ) from the datas obtained by making the vehicle to follow a series of previously generated

routes. The open loop PP did the work from the map of the environment of Fig6. The task has always been to reach a final point $\{q_f\}$ from different initial positions $\{q_0\}$ elected randomly. It is observed that the paths generated by the resulted PP, have good length and continuity properties (Table 1).

5. CONCLUSION

An architecture, and a method for the design of a dynamic path planner which utilizes fuzzy logic has been presented. The method consists in identifying a fuzzy mapping from the examples generated by an "expert": a model-based path planning. The dynamic path planner maps the ultrasonic sensors readings and the odometer measures to the robot control commands –its heading velocity and steering angle–. The correct mapping is found by identifying a set of fuzzy rules from desired input-output data pairs. The effectiveness of the proposed method is verified by a series of simulations. Quality metrics to compare the performance of different algorithms are introduced.

ACKNOWLEDGEMENTS

The research reported here has been partly supported by the *Fundación para el Fomento en Asturias de la Investigación Científica Aplicada y la Tecnología* (FYCIT), TAP92-0753.

REFERENCES

Bonarini, Andrea (1995). Learning to coordinate fuzzy behaviors for autonomous agents. *Tech.Report, Dip. di Elettronica e Informazione - Politecnico di Milano, 1995.*

Borenstein, J. and Y. Koren (1988). Obstacle avoidance with ultrasonic sensors. *IEEE Journal of Robotics and Automation, 4(2), Apr.1988.*

Brooks, Rodney A. (1986). A robust layered control system for a mobile robot. *IEEE Journal on Robotics and Automation, vol RA-2-1, March 1986.*

Dai Feng, Bruce H. Krogh (1990). Satisficing feedback strategies for local navigation of autonomous mobile robots. *IEEE Trans. Syst. Man Cybern., vol. 20, No. 6, Nov-Dic 1990.*

Gat, Erann (1993). On the role of theory in the control of autonomous mobile robots. *JPL Publication, Caltech Jet Propulsion Laboratory.*

Latombe, Jean-Claude (1991). *Robot Motion Planning.*

Lee, P. and L. Wang (1994). Collision avoidance by fuzzy logic control for automated guided vehicle navigation. *Journal of Robotic Systems, 11(8), pp.743-760, 1994.*

Lumelsky, V. J. and A. A. Stepanov (1986). Dynamic path planning for a mobile automaton with limited information on the environment. *IEEE Trans. Automatic Control, vol. 31, n 11, Nov 1986.*

Ollero, A. and A. García-Cerezo (1993). Design of fuzzy logic control systems.applications to robotics. *European Workshop on Ind. Fuzzy Control and Applicatios, Barcelona 21-23 Abril 1993.*

Saffioti, A., E. Ruspini and K. Konolige (1993). Robust execution of robot plans using fuzzy logic. *IJCAI Workshop on Fuzzy Logic.*

Salichs, M. A., E. A. Puente, D. Gachet and J. R. Pimentel (1993). Learning behavioral control by reinforcement for an autonomous mobile robot. *Proc. IECON'93 pp.1463-1441, Hawai Nov. 1993.*

Talluri, R. and J. K. Aggarwal (1992). Position estimation for an autonomous mobile robot in an outdoor environment. *IEEE Trans. on Robotics and Automation, 8(6), Oct.1992.*

Wang, Li-Xin (1994). *Adaptive Fuzzy Systems and Control.*

Wang, Li-Xin and J. M. Mendel (1992). Generating fuzzy rules by learning from examples. *IEEE Trans. Syst. Man Cybern, vol. 22, n 6, Nov-Dic 1992.*

Yen, John and Nathan Pfluger (1995). A fuzzy logic based extension to payton and rosenblatt command fusion method for mobile robot navigation. *IEEE Trans. Syst. Man Cybern., vol. 25, No. 6, June 1995.*

A FORWARD MARCH PATH-PLANNING ALGORITHM FOR
NONHOLONOMIC MOBILE ROBOT

ROHMER Serge

Systems Modelling & Dependability Laboratory
(LAEI, Metz University)
Troyes University
FRANCE

Abstract: Due to the kinematic constraints, mobile robots cannot follow an arbitrary path.
This paper describes a simple path-planning algorithm wich uses a forward march like
nonholonomic constraint. This method consists in creating for convex mobile robots, a set
of trajectories with the concept of "start window " and "image window". By definition, a
window is the frontier between two contiguous cells in a configuration space represented
with parallelepipeds. This algorithm perfectly adjusts to the problem of an electric
wheelchair for disabled person (VAHM project).

Keywords : nonholonomic, path planning, mobile robot, configuration space, obstacles
avoidance

1. INTRODUCTION

1.1 State of the problem

The specific problem studied in this paper is how to
plan a forwarding path for a mobile robot, knowing
its kinematic characteristics and its environment.
The proposed algorithm is an extension of a former
study concerning environment modelling, as well as
the discreet conceiving of a configuration space
thanks to the "Multivalue Numbers" (Pruski, 1990,
Pruski and Rohmer,1991; Rohmer, 1993). The
multivalue numbers represent the free space of an
environment with a set of rectangles called
"Continuous Multidimensional Multivalue
Numbers" NMMC.

The path planning of a mobile robot requires several
methods, whether we consider a local or global path
planning algorithm. This paper consider a global
path planning algorithm. Several methods approached
this problem by determining an admissible
configuration space, by reducing the robot to a point,
and growing the obstacles according to the
dimensional characteristics of the robot (Lozano-
Perez, 1983; Lozano-Perez and Wesley, 1979).
Moreover, the possibility of defining an exact
configuration space, makes it possible to certainly
determine holonomic trajectories from which

nonholonomic trajectories are deduced (Avnaim, et
al.,1988). The main constraint of the exact
determination of an admissible configuration space is
the algorithmic complexity and the lack of
malleability of these models. The method proposed
in this paper makes it possible to make up for this
problem. From the admissible space determination,
(Laumond, 1987) discovered that it was possible to
join two ordinary configurations in a connex
component, considering the kinematic characteristics
of the robot. These results guarantee a trajectory for
nonholonomic robots, but with a great algorithm
complexity. Another approach only based on a
geometrical reasoning, makes it possible to
determine a trajectory as a set of robot manoeuvres
taking into account the kinematic constraints
(Vasseur, 1992). And this, if the robots environment
in the form of convex polygons is considered.

1.2 A simple solution

The path planning problem for nonholonomic
mobile robots is tackled with a simple approach. In a
first part, an algorithm of forward path-planning is
described for mobile convex robot, from bases set up
by (Laumond, 1987) and (Siméon, 1989). In a second
part, the efficiency of the algorithm is showed to

advantage by describing the implementation in the frame of the VAHM project (Autonomous Vehicle for Disabled Person) .

2. NONHOLONOMY
2.1 *State of the problem*

Let's consider a robot moving thanks to two coaxial independent wheels. The robot is balanced by one of several swivel wheels. We define the configuration vector by 3 components $q = (x_R, y_R, \theta)$ with x_R and y_R the coordinates of the reference point of the robot and θ its orientation. The reference point is equidistant from the two driving wheels and located on their axis. In this case, the nonholonomic constraint is written as follows $\dot{y}_R = \dot{x}_R \tan\theta$, therefore the reference point of the robot can only move in the normal direction to the driving wheels axis. Thus, the characterisation of a system is essential to know if any two configurations can be connected by a way that can be done in a given time (notion of controllability). The research of (Barraquand and Latombe, 1989) made it possible to characterise nonholonomy from the integrability theorem of Frobenius, and this lead Latombe to define the system controllability. Laumond demonstrated the controllability of a robot like a car (Laumond, 1987).

The problem is to determine a nonholonomic path going through multivalue numbers corresponding to convex polygons. Siméon[9] proved it was possible to move in a parallelepiped from any point to an outlet window, and this, for every authorised orientation interval (Fig. 1). The resulting path is made of many segments which can theoretically be followed by a robot.

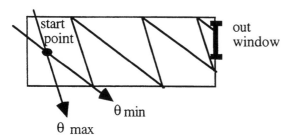

Fig. 1. Trajectory in a parallelepiped

In the case of a robot whose gyration radius is reduced, you only have to generate helixes (from arcs of circles) instead of segments.

In practice, the robot movement should always be controlled so that it follows the calculated path. However, because segments succeed very close together, the errors, accumulated during the movement prevent the path to be exactly followed. Moreover, orientation changes, small movements and the slowness resulting from such a path following, make the route very uncomfortable.

The trajectory can be improved by choosing parallelepipeds with maximal orientation intervals (decreasing of segments making up the trajectory if [qmin, qmax] is great). However, the resulting paths cannot guarantee the absence of manoeuvres. A manoeuvre is defined as a moving alternating forward march and reversing (turning back point). Solutions make it possible to significantly reduce manoeuvres, (Mirtrich and Canny, 1992) proposes to connect the source robot configuration to the final one with short paths made up with arcs of circles (Shortest feasible Path Ball). This solution is close to the one of (Taix, 1991) which "smoothes" a trajectory stemmed from a path planning for holonomic robot, with the Reeds and Shepp curves. These methods offer the advantage of giving comfortable trajectories, supposing that a holonomic path initially exists.

However, there is still a constraint : trajectories do not control the forward and reversing march.

The manoeuvres can be cancelled with a forward march path planning: the comfort is guaranteed.

2.2 *Forward march trajectory (Rohmer, 1993)*

A forward march trajectory is a set of succeed segments without turning back points for a nonholonomic mobile robot in a given environment. The path-planning algorithm determine a path from a start point to a target point with a method based on a concept of "in" and "out" windows in the configuration space.

Definitions

Let E be a free space in a configuration space represented with parallelepipeds (NMMC).
Let T_1 and T_2 be two NMMC of E, such as T_2 is closed to T_1 (Fig. 2).
• "in windows" called *iw*, is an area located on one side of T_1 . It represents the way that will be used by the path to go through T_1.
• "out window" called *ow*, is the area born from the contact or intersection of T_1 with T_2.
• "image window" called Im(*iw*) is the projection of iw on the sides of T_1 according to the authorised orientation interval.

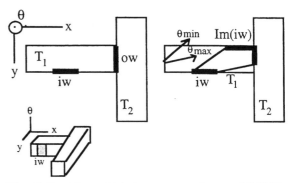

Fig. 2. Let T_1 and T_2 be two contiguous NMMC. Im(iw) is determined by the projection of iw according to the specific orientation interval $[q_{min}, q_{max}]$.

Proposal 1

Let T_1 and T_2 be two NMMC close to each other, iw in window in T_1 and ow its out window.

A segment T_a defines a forward march path from T_1 to T_2 if the intersection of $Im(iw)$ and ow is not empty :

$$\mathbf{if}\ Im(iw) \cap ow \neq \varnothing\ \mathbf{then}\ \exists\ P_1(x_1, y_1) \in iw$$
$$\mathbf{and}\ P_2(x_2, y_2) \in (Im(iw) \cap ow) \setminus [P_1, P_2] = T_a$$

Proposal 2

The part of $Im(iw)$ contained in ow becomes the iw of T_2. This part is memorized by a set of NMMC (Fig. 3) in order to be integrated in the multivalue tree for the path planning.

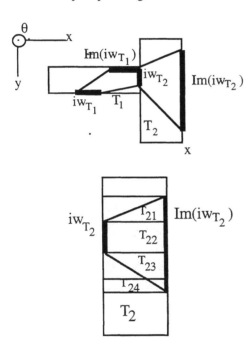

Fig. 3. Windows and NMMC images.
Let be T_{2i}, i NMMC born from $Im(iw_{T_2})$

The figures 4 and 5 show a simple example of path planning. A rectangular robot is in a corridor, and it has to reach a target located behind it. The path planning is computed with a classical algorithm for shortest way A*.

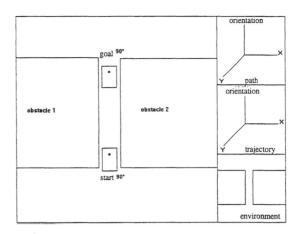

Fig. 4. Start and target configurations. The robot is located in a corridor, its start position and its final position have a the same orientation (90°).

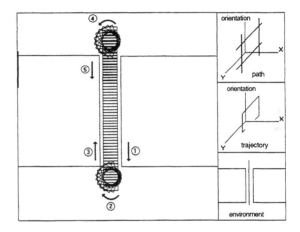

Fig. 5. Forward march path. The robot goes through the corridor (first step), makes a pure rotation (2nd step) and by its target point, goes through the corridor (3th step), makes a pure rotation (4th step), and at last, its target (5th step).

Let's have a look on the algorithm limits by detailing the advantages and disadvantages of a forward march path.

Disadvantages :

• Without the reversing march, every clearing manoeuvre is impossible. Therefore, several target configurations are impossible. The robot is uncontrollable.
• The obtained trajectory can be longer than the one in reversing march. Thus, the path-planning algorithm uses more nodes.

Advantages :

• The diminution of manoeuvres cancels short movements, and therefore decreases the number of errors avoiding the exact path following.

In the frame of the VAHM project for the automatization of an electric wheelchair for disabled person :
• This elimination of manoeuvres cancels slowness and uncomfort due to manoeuvres.
• It gives a maximal security to the disabled person. It allows this person to face the potential dangers, and therefore to control the vehicle movements at any time.

3. THE V.A.H.M. PROJECT (Bourhis, et al. 1993; Pruski and Bourhis 1992; Moumen and Pruski,1992)
3.1 *Introduction*

The V.A.H.M. project (Autonomous Vehicle for Disabled Person) is a study of a cooperation between a disabled person and mobile robotics. The purpose is to free the person of the moving constraints, thanks to an automatic system of navigation. The structure of the command system is made of three hierarchical levels : intelligent level, control level and executive level.
The research concerns the path planning for an electric wheelchair in an environment such as an apartment. In this frame, the case of a mobile base (ROBUTER) considered as an electric wheelchair is studied (Fig. 6).

Fig. 6. V. A. H. M.. A chair for the disabled person and a shelf with the computer are fixed on a mobile ROBUTER base (Robosoft). The computer has two roles: central computer and man-machine interface.

3.2 *The path planning*

The path-planning consists in determining the way of the wheelchair between a start point (origin of the wheelchair) to a target point (designated by the user in the wheelchair), while avoiding the obstacles.

The person in the wheelchair can automatically move towards a target point thanks to a man-machine interface made of a portable computer. After having designated the target point in a graphical environment, the path planning module determines the path to follow and sends this information to the path following module. This module receives the path sent by the central computer and commands the base movements as close as possible to the calculated path.
The resulting path generates the automatic movement of the wheelchair.

3.3 *Path planning tests*

A set of tests are set up with mobile base in real environments like rooms in a flat. The test proposed is a forward moving in an environment composed of 2 rooms linked by a corridor. The real dimensions of the rooms are converted(with the obstacles) so as to insert the whole environment in the modelling grid (256*256 cells) (Fig. 7). With this conversion, we obtain a 6.44 centimeters per cell accuracy (16.5 meters for 256 cells).

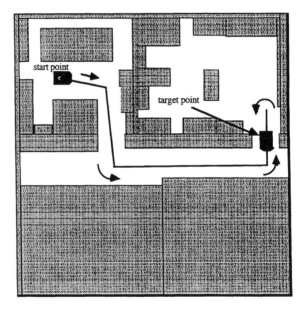

Fig. 7. Path.
   Calculation time for the path : 2.8 s
   Linear speed of the vehicle : 200 mm/s
   Angular speed of the vehicle : 0.4 rd/s

Once the target is designated and the path calculated, the coordinates of the segments are converted in order to be sent to the robot moving orders containing successively pure rotations and translations.

4. CONCLUSION

This paper shows how kinematic constraints of a mobile robot involve the development of special path planning algorithms. Uphill from this problem, it is necessary to create an admissible configuration space, knowing that an exact configuration space needs a very complex algorithm. Moreover, the conversion of such a space in the case of dynamic environment is not easy. A discreet configuration space used in an former paper allow to get round this problem.

The notion of "window" is introduced with the aim of creating a forward march path planning in a discreet configuration space.

The limits of the algorithm are very favourable to the VAHM project. However, if no forwarding path does exist, manoeuvres are necessary. These manoeuvres can be ordered by the user of the wheelchair, who is lead to reverse (potential dangers).

Thus, for a future research it seems necessary to use a path associating forward march for long distances with reverse march only if the vehicle is blocked, in order to guarantee the wheelchair controllability.

In a case of a very long path, the accumulation of moving errors can induce a gap between theoretical and real paths, thus it is necessary to take these errors into account. In a future paper, a method issued from equiprobability ellipses will be introduced in order to define a path predicting in advance the collision risks, so that a robust trajectory will be guaranteed.

REFERENCES

Avnaim, F., J.D. Boissonnat and B. Faverjon (1988). A practical exact motion planning algorithm for polygonal objects admist polygonal obstacles. In: *IEEE Conf. on Robotics and Automation*, Philadelphia, 1656-1661.

Barraquand, J. and J.C. Latombe (1989). On non-holonomic mobile robots and optimal manauvering. In: *Revue d'intelligence artificielle.*, vol.3, n°2, édition Herlmes, 77-103.

Bourhis, G. , K. Moumen, P. Pino, S. Rohmer, and A. Pruski (1993). Assisted Navigation for a Powered Wheelchair. In: *IEEE Syst. Man and Cybernetics Conf.* , Touquet, 553-558.

Laumond, J. P. (1987). Feasible trajectories for mobile robots with kinematic and environment constraints. In: *International Conference on Autonomous Systems*, Amsterdam, 346-354;

Lozano-Perez, T. and M. A. Wesley (1979). An algorithm for planning-free Paths Among Polyhedral Obstacles. In: *Communications of the ACM*, vol.22, n° 10, 560-570.

Lozano-Perez, T. (1983). Spatial planning: configuration space approach. In: *IEEE Transactions on Computers*, C-32, 108-120.

Mirtrich, B. and J. Canny (1992). Using Skeletons for nonholonomic path planning among obstacles. In : *IEEE Conference on Robotics and Automation*, Nice.

Moumen, K. and A. Pruski (1992). VAHM project : automatic control of mobility. In: *Robotics systems*, Kluwer Academic Publishers, 359-366.

Pruski, A. (1990).Multi-robot path planning among moving obstacles using mutivalue codes. In: *IEEE Conf on CIM*, Troy, 588-592.

Pruski, A. and S. Rohmer (1991). Multivalue coding: application to autonomous robot path planning with rotations. In : *IEEE Conf. on Robotics and Automation*, Sacramento, 694-699.

Pruski, A. and G. Bourhis (1992). The VAHM project: a cooperation between an autonomous mobile robot platform and a disable person. In: *IEEE Conf. on Robotics and Automation*, Nice, vol.2, 268-273.

Rohmer, S. (1993). Environment Modelling with Multivalue Codes; Robust path palnning for non-holonomic mobile robot. Thèse de l'université de Nancy I.

Siméon, T. (1989). Génération automatique de trajectoires sans collision et planification de taches de manipulation en robotique. Thèse de l'université Paul Sabatier, Toulouse.

Taix, M. (1991). Planification de mouvement pour robot mobile non-holonome. In: Thèse de l'Université Paul Sabatier, Toulouse.

Vasseur, H. A. (1992). Navigation car-like mobile robots in obstructed environnments using convex polygonal cells. In: *Robotics and Autonomous Systems*, vol. 10, 133-146.

MOBILE ROBOTS : DECISIONAL AND OPERATIONAL AUTONOMY

Georges Giralt

LAAS-CNRS
7, avenue du Colonel Roche
31077 Toulouse Cedex - France

Abstract : The field of Autonomous Mobile Robots has experienced an important growth, both considering basic research subjects related to Machine Intelligence as well as by the opening of a wide range of real world applications. We briefly describe three key generic concepts : autonomy, reactivity and human-machine intelligence that support seminal contending paradigms for autonomous robot systems. Two large case studies, a multi-robot system for cargo trans-shipment and a rover for planetary exploration, illustrate the central ideas and provide with a precise background for on-going research.

Keywords: Autonomous mobile robots, machine intelligence, decisional and operation autonomy.

1. INTRODUCTION

During the last decade, the field of Autonomous Mobile Robots has experienced an important growth, both considering very basic research subjects aimed to develop the set of enabling technologies with a strong emphasis in Machine Intelligence as well as by the opening of a wide range of real world applications (Giralt, 1992a; Giralt 1993).

Let us briefly discuss first the basic research issues. They cover three vast domains :

i) **Automated locomotion** which encompasses the design of mechanical structures ranging from wheels to a variety of all-terrain devices that may combine tracks, legs,...

Sophisticated control systems have consequently to be divised, able to cope with kinematic and dynamic motion characteristics, that may include 3D terrain constraints as well as ground deformation.

ii) **Artificial perception** which includes a highly challenging class of subjects ranging from natural scene analysis, multi-sensor fusion, 3D environment modelling, localization, and on-line sensory feedback.

iii) **Planning, decisional and operational autonomy** which entail and support today probably one of the most exciting domains for research and debate in Machine Intelligence that may be encapsuled within the interplay of such concepts as Autonomy, Reactivity, Programming. Here is the real challenge : how to integrate in an efficient way reasoning and task planning, task level programming, real time systems, and sensor based execution control.

The assessment of the current development of Advanced Robotics, supported by information technology, microtechnologies and microsystems developments, has lead to consider for Robotics the new frontier of non-manufacturing applications. Indeed, sensor-based systems and embedded intelligence are decisive factors for a host of new machinery and products that will take the robot out of the floorshop :

- to professional service domains : surveillance, plant maintenance, public safety, rescue and disaster prevention,...

- in field-based applications such as mining, civil engineering, forestry, agriculture, underwater, exploration/science (planets, polar bases,...)

- to public-oriented areas ranging from domestic and professional cleaning to assistance to the disabled and the aging

The variety, the very broad scope encompassed, stress the importance and the far-reaching possibilities of the overall domain.

We should note that the above applications range from very cooperative (surveillance) to truly non-cooperative environments (planet, e.g. Mars exploration).

In this paper, we consider the class of task-level programmable mobile-robots in relation to the broad set of applications that include tasks not entirely pre-defined at design stage to be executed in an environment that has not been completely engineered to cooperate with the robot.

As a consequence of the lack of cooperation between the task environment and the machines, we can infer a key requirement for this class of Mobile Robots : since they have to adapt in real time to the task, it is necessary to include **on-line** human and/or machine intelligence.

In the following sections, we first discuss the generic concepts which are today important research streams and mature enough in their development to be efficiently implemented in real world applications.

Next, we will illustrate the real world possibilities by describing the features of two large application projects we are currently working in at LAAS and that we believe to be significant and demonstrative:

- a multi-robot system for cargo trans-shipment, representative of service robots for partially-structured environments,
- a rover for planetary exploration, representative of Field Robots applications.

2. GENERAL CONCEPTS

2.1. Autonomy

The key factor in automation in general, and above all in robotics, is how closer to the commands or the objectives that were set to the system, are the responses obtained, i.e. how much the evolution of the system can be predicted.

There are cases of generic predictability for the tasks that are compatible today with the state of the art in robotics. There are also a lot which are **not** compatible, e.g. it will utterly be science-fiction to send a machine to look for an "interesting rock" during a planet exploration, without further description, or to assembly a given set of equipment without step by step instructions. However, it is clearly possible to contend that an efficient

automation can be achieved, that may combine different levels of cooperation between remote machines and human operators, the levels of robotics sophistication and the necessary on-board machine intelligence depending on the task to be achieved.

Robot autonomy represents today the leading edge of advanced robotics research. The variety of tasks to be performed by the robot range from object manipulation to machine mobility, within a world environment which may be largely unknown and/or not static.

Therefore, the basic research support concerns mobile robots that have to carry on tasks in environments that cannot be engineered to perfectly cooperate with the machine, and thus have to exhibit some level of Machine Intelligence. Several important issues are consequently brought up, such as the levels of modelling to be used, both for the task and the work environment, the computational system structure, the operating functionalities that we want to obtain, including the capacity -or the absence of this capacity- of operator control and programming modes, and the sensor-based execution control systems.

2.2. Reactivity

Reactivity plays a central role in Robotics, as well as in some fields of AI such as planning that can be viewed as only partly related to Robotics.

This key concept is a salient attribute that has to be implemented at the very core of any real world operating robot. In fact, reactivity is the feature which connects the actual state as perceived by the sensors, and the implicit or explicit commands that the machine has been set to execute. This is true for early architectures with simple feedback loops as well as for machine- intelligence embedding systems ranging from Behavior based ones to functional hierarchical systems with planning capacities (see Section 3).

To clearly understand the issues at hand amidst the variety of system approaches, reactivity has to be considered at three levels :

(i) reflex actions and servoing feedback processes,
(ii) decisional and reasoning processes implying context-dependent reactions to signals and data inputs,
(iii) task and environment changes that may imply task replanning.

2.3. Human-Machine Intelligence

We have contended that decisional and operational autonomy is today a feasible technique that can be used at several levels to make more efficient and pleasant to operate a machine that can be either local or possibly located in a remote and hostile site.

The person-machine interplay has to be considered within different classes :

- Telerobotics which implies a human-machine interaction in which the operator directly controls the machine. Machine-intelligence is here implemented both on board and as a system support for the operator. A broad avenue of developments is here opened by technics such a Virtual Reality or much more appropriate Augmented Reality Systems.

- Task-level programmable robots which implies higher degrees of decisional and operational autonomy implemented on board and retains the use of machine-intelligence at the operator level to support task specification and task execution supervision.

Let us now consider the case in which humans and machines have to permanently interact and to cooperate within an integrated system that includes both. This is of course the case in vehicles such as computer assisted planes and cars, where the driver/pilot may use some level of machine intelligence as a support but remains in control. A very difficult and exciting problem arises here. It streams from the concept of co-autonomy. Here **both** human and machine might be endowed with funtionalities corresponding to the highest control level.

The functional modalities must confront and properly handle a crucial two folded problem.The machine must obey the user and take care of details in executing the actions. It must also decide to take over control for security reasons, while yielding to the user as necessary. It must issue warnings and explain its decisions if needed (Chatila, et *al..*, 1995).

3. OPERATIONAL AUTONOMOUS ROBOT CONCEPTS

There are a broad variety of cases ranging from tasks where it is requested not to burden an operator, to work in unstructured, ill-known or possibly entirely unknown environments, that clearly imply real time, intelligent adaptation, which cannot be fulfilled efficiently by on-line programming from a remote system.

This, for an Intervention Robot, means the need for the machine to possess **operational** and **decisional autonomy**.

Although strictly speaking autonomy can as well be attached to manipulation, research on most of the basic issues is truly at home with mobility.

Historically, everything began in 1969 at SRI with Shakey (Fikes, *et al.*, 1972), the first autonomous mobile robot project. HILARE, the most senior mobile robot "still running" was launched in 1977 (Giralt, *et al.*, 1979).

Among the variety of concepts, architectures and systems, we believe that two paradigms stand out as contenders to embed on-board **machine intelligence**, i.e. decisional and operational autonomy, at a meaningful level :

- Autonomous Behaviour based machines.
- Autonomous task level programmable machines.

3.1. Autonomous behaviour-based machines

During the eighties, emerged with R. Brooks as the proponent, the concept of machine intelligence for mobile robots based on a processing architecture wherein :

- a set of reactive modules, structured in sub-systems **Behaviors**, directly process sensory data to produce output actions,

- the Behaviors are integrated in a layered, hierarchical structure. A higher level subsumes a lower level by interfering with its inputs/outputs (figure 1) (Brooks, 1986).

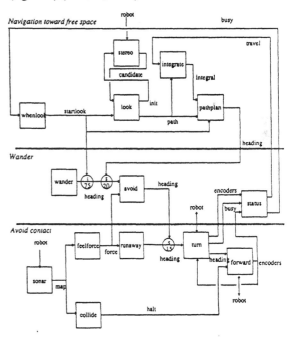

Fig. 1 An example of layered, behavior based structure.

Consequently, the robots built accordingly evolve in an entirely autonomous way, autonomy meaning at its very basic level total **absence of control** and of programming/supervision link with an operator station.

Accordingly, we will characterize this concept on the basis of two salient features :

- elementary "**Behavior**" levels, i.e. simple sensing-action closed-loop processes,
- a **wired** decisional structure embedding the reactive processes, i.e. intelligence and **autonomy** being established at **design level**.

Behavior-based architectures have flourished and systems have been developed in many places, but not always integrated in the architecture proposed by MIT.

3.2. Autonomous task level programmable machines

In contrast with the concept to achieve autonomy by the sole means of a set of reactive processes, work on operational and decisional autonomy for mobile robots involving levels of explicit reasoning, which can be viewed as the implementation of **deliberative intelligence**, is actively pursued and very particularly with space exploration objectives.

We would like to mention two projects in this context, because they underline well the system approach : the United States project AMBLER at CMU (Simmons, 1992), and the French project VAP (see section 4.2).

In the following, we will specially refer to what we think a very representative and comprehensive system of the class of autonomous intervention mobile robots, the Remote Operated Autonomous Robot concept, developed at LAAS in the framework of the HILARE project.

In this approach, the robot is a programmable machine and the attributes attached to the term autonomous can be translated into the concept of "Task Level Programmable Robots" in which the elementary operators and language primitives include a set of **sensor-driven elementary tasks** (Chatila, *et al.*, 1991; Giralt, *et al.*., 1992).

Furthermore, the supplementary capacities and attributes at operator level are devised to compensate for unrealistic assumptions on Machine Intelligence and to achieve an efficient overall system wherein the robot has the best possible level of operational and decisional autonomy, while remaining under the control of the operator at a supervision and programming level, i.e. the robot is a **task-level programmable machine**.

The concept is implemented into a functional architecture which is composed of two systems (figure 2) :

- The operator station, which includes modules and facilities for task level programming, mission planning, data bases and environment model updating,... Given a set of goals (mission) to be achieved, the operation station plans and then refines the mission, generating a set of tasks, to be interpreted and performed by the robot, along with their execution "modalities".

- The robot control system which possesses all the functional capabilities of an autonomous robot :

. a supervisor plus a task refinement planner, which further refines the tasks, i.e. produces plans of actions, according to the actual state of the robot and its environment

. an Executive which organizes and manages the activities of the robot that are fulfilled by

. a set of functional modules that control the robot's sensors and effectors and provide for basic algorithmic capacities.

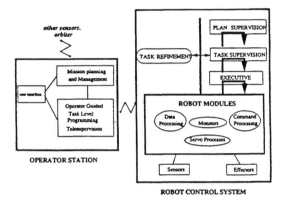

Fig. 2 Remote-Operated Autonomous-Robot functional architecture

The robot is **fully autonomous** at task level. It receives tasks assignments that it transforms into sequences of actions using its own interpretation and planning capacities, and executes these actions while being **reactive to asynchronous events** and environmental conditions.

The salient features of this concept are :

- **reactive** and **deliberative** intelligence (i.e. task **planning**, **reasoning**) are integrated in the on-board systems
- **task level programming**, i.e.. symbolic abstract off-line programming, which allows for robot control in the nominal mode whenever necessary.

4. CASE STUDIES

4.1. A multi-robot system for cargo trans-shipment

Let us consider the generic case for service/intervention robots for partially structured environment where we want to operate a large fleet of autonomous robots in a route network composed of lanes, crossings and open areas (Alami, *et al.*, 1995a).

Typical applications of this problem are load transshipment as dealt with in the MARTHA project[1] which requires the development of a large fleet of autonomous mobile robots for the transportation of containers in harbors, airports and railway station environments.

4.1.1. MARTHA : project outline

Two testbeds are considered for validation and demonstration of the project results :

- A heavy-load sea-rail trans-shipment operation
- A medium-load airport transportation and handling operation

Regarding the first testbed, it is important to mention that the Rotterdam harbour (Europe Combined Terminals) offers already an operational system equipped with about fifty unmanned vehicles for container transportation between automated pier cranes and storage stations.

The operating environment is both indoor or outdoor (all weather), partially structured (because evolution areas may be often reconfigured) and dynamic (because of the robot motion and of the presence of other active agents like humans or carts), hence a global localization and new sensing capabilities are necessary. Improved autonomous navigation performances are also needed so that robots can perform in a satisfactory way such special behaviours as mobile obstacle avoidance, crossing, convoy driving, whenever it is required and without having to be directed by the central station (figure 3).

The solutions proposed by MARTHA rely on two main concepts :

- Overall task planning is handled by the central station which provides appropriate software support to the operator machine interfaces for task assignment to every robots. At this level, only general information about nominal and alternate possible routes and time constraints are provided to the machines. The operating environment is a route network composed of entities like lanes, crossings and open areas.

- On-board decisional autonomy based on multi-sensory perception system and decisional software allowing the machine to plan for all local actions including trajectory definition and multi-robot interaction such as crossing and overtaking.

---

[1] MARTHA : European ESPRIT Project n° 6668. "Multiple Autonomous Robots for Transport and Handling Application"

This two-folded approach is highly demonstrative of the benefits that can be obtained to implement in an efficient and reasonable way the two central concepts of central station task-assignment and distributed machine intelligence embedded into sensor-based machines.

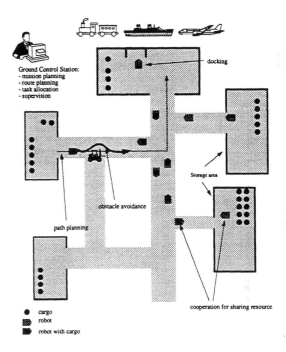

Fig. 3 Cargo-trans-shipment yard layout

4.1.2. Decentralized cooperation of autonomous robots (Alami, *et al.*, 1995b).

The Plan Merging Paradigm

In order to handle the problem of the decentralized cooperation of a fleet of autonomous robots, a plan-merging paradigm has been proposed which is well suited to such applications as it allows to deal with a great number of robots, locally dealing with conflicts while maintaining a global coherence of the system. Indeed, it limits the role of the central system to the assignment of tasks and routes to the robots (without specifying any trajectory or any synchronization between robots) taking only into account global traffic constraints.

The Environmental Model

In order to allow efficient and incremental plan merging, the route network is decomposed into smaller entities called "cells" or "spatial resources" which will be used as a basis for dealing with local conflicts.

Basically, the robots navigate through an oriented graph of cells.

Lanes and crossings are composed of a set of connected cells, while areas consist of only one cell.

Thus, the environment model, which is provided to each robot, mainly contains topological and geometric information :

- A network describing the connections of areas and crossings
- A lower level topological description (cell level)
- A geometrical description of cells (polygonal regions)
- Additional information concerning landmarks (for re-localization), station descriptions for docking and load handling actions.

The Central Station

The central station is in charge of producing the high level plans to load/unload ships/trains/planes (in response to user request). The plan produced takes into account the topological model of the environment as well as the availability of such or such robot. However, it does not further specify the sequence of robots going through a crossing (this decision is left to the robot locally concerned), nor does it require the robot to remain on the specified lanes (in case it needs to move away from an unexpected obstacle).

A Plan-Merging Protocol for Multi-Robot Navigation

The cooperation scheme used is the general paradigm, called Plan-Merging Paradigm where robots incrementally merge their plans into a set of already coordinated plans. This is done through exchange of information about their current state and their future actions.

For the case of a number of mobile robots in a route network environment, a specific Plan-Merging Protocol has been divided based on spatial resource allocation but in this context, Plan-Merging Operation is done for a limited list of required resources (a set of cells which will be traversed during plan merging).

Plan-Merging for cell occupation

In most situations, robot navigation and the associated Plan-merging procedure are performed by trying to maintain each cell of the environment occupied by at most one robot. This allows the robots to plan their trajectories independently, to compute the set of cells they will cross and to perform Plan-Merging at cell allocation level.

It is very efficient to choose an allocation strategy which makes the robots allocate one cell ahead when they move along lanes, while for crossing,

they must allocate all the cells necessary to traverse and leave it. This is done in order to not to constrain unnecessarily the other robots.

4.2. A rover for planetary exploration

The French Space Agency (CNES) undertook in 1989 a study project, VAP (Automated Planetary Vehicle) aiming to assess the state of the art and to develop the base line system concepts for Planetary Rovers. In the first phase of the project, October 1989 to October 1993, the corresponding robotics studies have been carried out by CNES in association with RISP[2], a Consortium of Laboratories belonging to four R&D organizations (CEA, CNRS, INRIA, ONERA).

Various Lunar and Martian missions have been considered with a main thrust on an ambitious scientific target mission on Mars by the end of the millennium. This last program requires a large planet traverse of the order of 1000 kilometers, over the 13 months sand-storm free period of time in a Martian year.

A realistic approach imposes a thorough assessment of the various constraints to be met :

(i) Terrain constraints : the surface of Mars consists of mountains, canyons and valleys, craters, blocks, rocks, and sand dunes, low-coherence drift material.

(ii) Resources and system constraints : Volume, mass, energy - Space qualified equipment - Radio transmission constraints (up to 20 minutes, time delay, severe bandwidth limitations),...

The proposed global VAP concept stresses two essential aspects for a machine to qualify by the end of the millennium for a Mars exploration and scientific experiments mission :

- the capacity to be assigned different tasks by an Earth-based operator (task-level teleprogramming), and

- the necessity for the rover to exhibit operational and decisional autonomy.

The functionalities and the system components which correspond to this concept and which take into account the environmental and system constraints have been defined, as well as the functional architecture.

Following the research programme, it was decided in 1993 to build a demonstration rover to validate the robotics functionalities and to experimentally demonstrate the degree of operational and decisional autonomy offered by concepts developed since 1989,

---

[2] RISP : Intervention Robots for Planetary Exploration

at the core of which the backbone functional architecture that has been described in section 3.2.

The Eureka pan-European structure appeared to offer a logical continuation to the French studies while also seeking to federate and transfer to the field of space expertise and achievements generated by other programmes in mobile robotics in which both industry and research participate.

The objective of the IARES project (Boissier, *et al.*, 1994) is to build a ground demonstrator that will be fully representative of robot functionalities and of the autonomy required for planetary rover and therefore to develop the techniques that will allow the exploration and subsequent operation on planetary sites (Mars, the Moon). The following aspects relating to mobile robotics are developed :

- algorithms and systems for travel over natural and ill-known terrain,
- equipment required to autonomously determine the robot's position,
- equipment required for autonomous on-board environment perception,
- aspects related to on-board decision-making and the planning of operations associated with the operator station on Earth,
- handling and gripping tools.

Therefore, the objectives of the IARES demonstrator (figure 4) include both the definition and the test of all the sub-systems as well as the integrated structure in all operational modes with particular emphasis on the nominal one exhibiting on-board decisional and operational autonomy associated with the mission and task-level programming capacities from the operator station. IARES will be in particular tested on the GEROMS test site at CNES in Toulouse (France).

IARES associates 9 partners from 4 countries : France : ALCATEL ESPACE, CYBERNETIX, ITMI, MATRA MARCONI SPACE, SAGEM, RISP ; Spain: IKERLAN ; Hungary : KFKI/RMKI; Russia : VNIITRANSMASH.

Since early 1993, VAP and IARES have benefited from a proof-of-the-concept demonstration, carried on by LAAS with the rover ADAM[3] in the on-going experiment EDEN (Chatila, *et al.*, 1994).

The aim of Eden is to demonstrate a fully autonomous navigation in a natural environment. The canonical task is "GO-TO (Landmark)" in an a priori unknown environment that is gradually discovered by the robot. The landmark is a known object given by a model. The on-going work includes perception, environment modeling,

---

[3] ADAM is a six drive-wheels platform propriety of Framatome and Matra lent to LAAS.

localization, path and trajectory planning and execution on flat or uneven terrain.

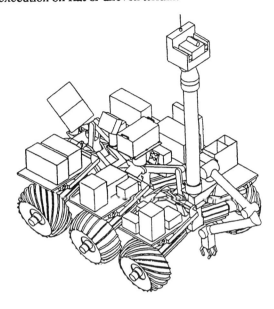

Fig. 4 IARES ground demonstrator concept

The generic control architecture for autonomous mobile robots developed at LAAS, organized into three levels, is instantiated in the case of the EDEN experiment.

At the top end, the higher task planning level plans the mission specified by the operator in terms of tasks, with temporal constraints interpretable by the robot and monitors mission execution. This level is not currently used in the experiment. It will be embedded in the teleprogramming environment application dedicated experiments such as planet exploration.

At the bottom end is the ``Functional Level'' that embeds the functions, encapsulated in modules, related to data acquisition and processing, environment representation, trajectory planning and motion control.

In between is the ``Decisional Level'', implemented in PRS (Ingrand *et al.*, 1992). It plans by refining the mission tasks, the actions involved in the task - here navigation - and controls their real-time execution according to the context and the robot state to achieve them. Functionally speaking, it is composed of a task planner (that selects the adequate procedures), a supervisor (that establishes the dependencies between the modules, combining the functions according to the context and modalities at run-time), and an executive that is actually the interface with the functional level to which it sends requests and checks the coherence of the various concurrent primitives.

Determination of the navigation modes is based on a fast analysis of the raw 3D data produced either by a Laser Range-Finder or by a stereovision algorithm. This analysis provides a description of the terrain in terms of five classes : [horizontal, flat with slope, uneven, obstacle, unknown].

The general approach is the following :
- perception (vision) detects and estimates the position of the landmark (goal) ;
- simple algorithms are first used to characterize the terrain into five classes. Uneven terrain regions are modelled more accurately with appropriate data structures (elevation map, B-splines,...) if the robot has to move through such regions ;
- a global navigation planner selects a subgoal within the known part of the environment ;
- the adequate motion planner (2D or 3D) is selected according to the nature of the traversed terrain (free and flat, cluttered and flat, uneven).

5. CONCLUSION

Mobile robots endowed with enough on-board machine intelligence are today in the front line of advanced research in machine intelligence. The kind of systems that provide them with decisional and operational autonomy are generic and can be used as intelligent modules in a very broad set of application domains implying manned or teleoperated vehicles.

In the paper, we have chosen to illustrate some key concepts such as autonomy, reactivity and human-machine intelligence, and the subsequent Task-level Programmable Robot functional architecture, by two case studies carried on at LAAS by the Robotics and Artificial Intelligence Group. The first one, a multi-robot system for cargo trans-shipment (section 4.1) is a very good example of the domain implying partially structured environments. The second, a rover for planetary exploration (section 4.2) certainly represents the best possible vector in the highly promising domain of Field Robotics. This project, besides the direct impact for hostile sites exploration such as Antarctica, Mars and the Moon, develops all the generic technologies for a host of socially and economically relevant applications (forestry, mining,...).

Finally, we have pointed out the novel and most demanding field of public oriented applications among which, indeed, personal robots to assist the impaired and the aging represent the highest challenge, both technically and by its economical and social impact.

REFERENCES

Alami R., F. Robert, F. Ingrand, S. Suzuki (1995a). Multi-robot cooperation through incremental plan-merging. *IEEE ICRA'95*, Nagoya (Japan), May 1995.

Alami, R., L. Aguilar, H. Bullata, S. Fleury, M. Herrb, F. Ingrand, M. Khatib, F. Robert (1995b). A General framework of multi-robot cooperation and its implementation on a set of three Hilare robots. In : *Fourth International Symposium on Experimental Robotics, ISER'95*, Stanford, California, June 30-July 2, 1995.

Boissier L., L. Petitjean (1994). System design and architecture of the IARES mobile testbed project. *Second IARP Workshop on Robotics in Space*, Montreal, Canada, July 6-8, 1994.

Brooks, R.A. (1986). A robust layered control system for a mobile robot. In : *IEEE Journal of Robotics and Automation*. April 1986.

Chatila R., R. Alami , G. Giralt (1991).Task-level programmable intervention autonomous robots. In: *Mechatronics and Robotics 1*. P.A. Mac-Conaill, P. Drews and K.H. Robrock Eds. IOS Press, pages 77-87, 1991.

Chatila R., R. Alami, S. Lacroix, J. Perret, C.Proust (1994). How to Explore a planet with a mobile robot ? *Second IARP Workshop on Robotics in Space*. Montreal, Canada, July 6-8, 1994.

Chatila R., P. Moutarlier, G. Giralt (1995). Personal robots to assist the impaired and the aging. *First IARP Workshop on Robotics for the Service Industries*. Sydney, Australia, 18-19 May 1995.

Fikes R.E., P. Hart, N.J. Nilsson (1972). Learning and executing generalized robot plans. *Artificial Intelligence*.

Giralt G., R. Sobek, R. Chatila (1979). A multi-level planning and navigation system for a mobile robot. A first approach to Hilare. *Six International Joint Conference on Artificial Intelligence*. Tokyo, Japan, August 20-24, 1979.

Giralt G. (1992a). Intervention mobile robots : generic concepts and main application domains. *23rd International Symposium on Industrial Robts*. Barcelona, Spain. 6-9 October 1992.

Giralt G., L. Boissier (1992b). The French planetary rover VAP : concept and current developments. *IROS'92*, Raleigh, USA, July 7-10, 1992.

Giralt G. (1993). Outdoor mobile robots. *International Conference on Robots for Competitive Industries, Brisbane, Australia, July 14-16, 1993*.

Ingrand F.F., M. P. Georgeff, A. S. Rao (1992). An Architecture for Real-Time Reasoning and System Control. *IEEE Expert*, Knowledge-Based Diagnosis in Process Engineering, 7(6):34--44, December 1992.

Simmons R.G. (1992). Monitoring and error recovery for autonomous walking. *IROS'92*, Raleigh, USA. July 7-10, 1992.

An Optical Localization System for
Mobile Robots Based on Active Beacons

F. Giuffrida, P. Morasso, G. Vercelli, R. Zaccaria
DIST, University of Genova, Via Opera Pia 13a, Genova, Italy
e-mail: sand@laboratorium.dist.unige.it

Abstract. The standard method for controlling the trajectories of AGVs in industrial applications
is based on wire-guide systems. It is robust but it lacks a flexible and economical localization system
which is essential for the development of intelligent navigation capabilities and the evolution of AGVs
to cooperating autonomous agents. This paper describes an active localization system developed for
indoor applications that allows an accurate localization of mobile vehicles in static and dynamic
operation. Simulations and experimental results with a prototype implementation are described.

Key Words. Autonomous mobile robots, navigation, localization, shop-floor oriented systems

1. WHY LOCALIZE?

Navigation techniques are usually divided into
three main classes: i wire-guidance (WG), ii reference landmarks guidance (LG), and iii dead reckoning (DR). The three classes can be ranked according to different criteria, such as positioning
accuracy, reliability, required computational resources, speed, flexibility, environmental impact.
The ideal robotic navigation system, in analogy
with biological counterparts, should be characterized by top performance for all the different
criteria, i.e. be very accurate, reliable, flexible, quick, computationally efficient, and environmentally non-invasive. Unfortunately, no single
navigation technology presently available can approach such goal.

• WG, although presently the most common solution in industrial applications, is terrible from
the point of view of flexibility and environmental
impact.

• For DR, the main drawback is that the positioning error cannot be bounded but is a growing
function of time.

• As regards LG there are at least three different
variations: in one version (LG_V) the landmarks are natural visual features of the environments;
in the second one (LG_P) the landmarks are special patterns, inserted in known positions of the
environment, which require some kind of pattern
recognition because are *passive* with respect to the
robot; in the third one (LG_A), the landmarks are
active in the sense that react to the interrogation
of the robot according to some kind of communication protocol. LG_V is the best one as regard-

s environmental impact, but the dependence on
computer vision and pattern recognition make it
slow, unreliable and computationally expensive,
particularly in practical industrial applications.
LG_P is an attempt to simplify the computer vision
task but still suffers from the general difficulty of
current vision systems in real environments, away
from the controlled variability of the labs. Vision-based techniques allow, in general, to compute the
position of the robot from a fixed relation with the
position and attitude of the camera mounted on
the robot itself: *model based perspective inversion*
is the classical method to recover the 3D position of the camera (Garibotto and Masciangelo,
1992). In LG_A the landmarks are *active beacon*s thus substituting the vision/pattern recognition
task with a communication protocol and a triangulation procedure, at the expense of a little more
environmental impact, much smaller than for WG
in any case.

Among the different aspects of the navigation
task, the *localization problem* is the crucial one
and consists of maintaining in real-time a reliable estimate of the position of one or more mobile robots with respect to a reference frame in
the environment. It is somehow surprising that
this problem is still an open issue in the current robotic literature: in spite of the wide use of
localization systems in the aeronautic and marine areas, only in the very last years the interest of industrial robotic companies has grown
up, with the introduction of ultrasonic, infrared
(G.L. Miller, 1990) and laser technology for indoor(C.D.McGillem, 1988), and GPS (global po-

sitioning system) for outdoor mobile robot applications(B.W. Parkinson, n.d.). The navigation system of an autonomous mobile robot needs a periodic position and orientation estimation, using external sensors, in order to reset drifts of its own odometry, introduced by wheel-slippage and ground deformation. In order to avoid the computational burden of vision-based approaches in recognizing natural landmarks, *artificial beacons* have been introduced in many approaches to solve in a simpler way the localization problem; for example, special geometric patterns (as circular rings) are extensively used to recognize relevant places in static museal environments (Masciangelo *et al.*, 1994). Other approaches use ultrasonic, infrared and laser sensors, as well as accelerometers (as the Inertial Navigation System (Miller *et al.*, 1989)). Moreover, combinations and integration of technologies are well suited to reduce uncertainties in estimates by means of some kind of Kalman filter-based algorithms, as in (Crowley, 1989). When using infrared and laser technologies instead of cameras or ultrasonic range finders, the artificial beacons can be either active or passive. Active beacons are essentially simple communication devices which exchange specific patterns with the transmitter/receiver onboard the robot. Passive beacons, on the contrary, are usually bar codes or retroflectors, which guarantee simplicity and economy, but are not sufficiently reliable in industrial, 'dirty', shop floors.

Our approach is a mixture of LG_A and DR. The latter gives a continuous sensory information and is used on a short-time basis, whereas the former one gives an intermittent but accurate position information. The positioning technology uses a modulated optical system (laser+infrared, patent pending) with active beacons in the working environment and a rotating unit mounted on the mobile robot. The measuring principle has some similarity with GPS, but instead of measuring the time delay and direction of arrival of electromagnetic waveforms from satellites to the on-board receiver (TDOA), our system measures angular delays and directions of arrival of laser beams (ADOBA: angular difference of beams arrivals).

2. STRUCTURE OF DYNAMIC LOCAL POSITIONING SYSTEM

DLPS is a positioning system for indoor applications in multi robot-environment, using modulated light beams (red and infrared). The system is composed of two parts: an on board **rotating unit** and a set of **active beacons** distributed over the area. Both units have transmitting/receiving elements and a communication processor, which allows to pack/unpack bit serial messages. The

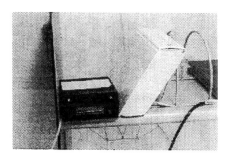

Fig. 1. Active beacon

Fig. 2. Rotating unit

rotating unit (see figure 2) transmits a rotating, modulated beam of light (red laser) and has an omnidirectional infrared receiver. A rotating mirror, powered by a DC motor, is used to change the direction of the laser beam and an encoder measures the current angular direction of the beam with respect to the mobile base. The rotation axis is vertical and the scanning plane is horizontal approximately at 2 meters from the ground in order to minimize the interactions of the rotating beam with humans and/or inanimate obstacles. The vertical aperture of the beam (about 10°), obtained by means of a cylindrical lens, is necessary in order to recover alignment problems and allow the beam to reach all the visible beacons. The frequency of rotation varies from 0.5 hz to 2.0 hz, depending on the area of operation. The system operates also as a transmitting unit. an higher rotational velocity allows more localizations. but the time the laser beam is seen from the receiver is reduced, limiting the information amount to transmit.

The **active beacons** (see figure 1) are little boxes without moving parts: both the receiver and the transmitter have a wide angle of view. The working principle is the following:
(ı) The rotating beam emitted by the robot is modulated in stream mode, carrying the identification code of the mobile base and the angle measured by the encoder connected to the rotating mirror;

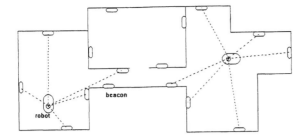

Fig. 3. Example of distribution of beacons in a working environment with two operating robots.

(ii) Each activated beacon answers back, adding to the received message its identification number; (iii) The answerback messages, collected by the mobile base for a full rotation of the mirror, are processed by the algorithms for the static and dynamic reconstruction of the position and the orientation of the robot.

This system differs from other beacon based localization systems because the beacons are not passive optical reflectors, but active units. Using the active beacons technique, the reliable identification of beacon and robot signals allows a unique reconstruction of each robot position. even in a multi-robot environment, provided that each robot has an explicit map of the active beacons in the environment, with their coordinates (see figure 3). Moreover, as each beacon has a processor unit, information on the operations made by the robots can be passed to host computers directly through the wired or optical links.

3. LOCALIZATION ALGORITHMS

The localization algorithms are used to control and correct the position computed by the odometry system: the available data for the localization module is a list of triples *[(beacon identifier), (beacon coordinates), (viewing angle)]*, which correspond to intermittent measures, and continuous measures coming from odometry. The problem is to fuse them in the same computational process. In particular, we distinguished two different algorithms:

- Static Localization: the position of the stationary vehicle in the environment is estimated using standard triangulation algorithms. This gives the most accurate measure.
- Dynamic Localization: the position of the moving vehicle is estimated in an approximated way by merging the intermittent optometric measures with the continuous odometric measures. As speed increases, the relative weight of the odometric part tends to increase and the global accuracy tends to decrease.

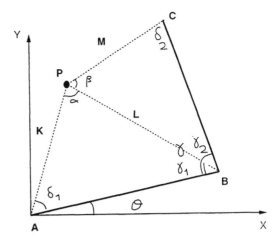

Fig. 4. Triangulation method.

Starting from the geometrical solution of the problem of inverse intersection triangulation, it is possible to solve the localization problem, known the beacons position in the area. The solution is based on the inverse triangulation Snellius theorem, which is a well known technique in topographical measuring applications.

Given 3 beacons, say A,B,C (see figure 4), of which we know the coordinates with respect to a reference system fixed in the environment: $A(x,y)$, $B(x,y)$, $C(x,y)$. The robot is identified by the position of a reference point P and the orientation $\vec{V}$ of its longitudinal axis. The localization algorithm computes P and $\vec{V}$ by using the angles measured between the robot position P and the beacons A, B, C.

For sake of simplicity, we suppose the origin O of the 2D space coincident with beacon A, and the direction of the robot $\vec{V}$ coincident with the vector $\vec{PA}$. Firstly, we define the following quantities:

$$\alpha = A\hat{P}B \quad \beta = B\hat{P}C \quad \gamma = A\hat{B}C \tag{1}$$

$$\delta_1 = P\hat{A}B \quad \delta_2 = P\hat{C}B \tag{2}$$

$$\eta = \delta_1 + \delta_2 = 2\pi - \gamma - \alpha - \beta \tag{3}$$

where α, β are measured at run-time. and γ is measured at calibration time. Applying sinus theorem to APB and BPC triangles we have:

$$\frac{\bar{AB}}{\sin(\alpha)} = \frac{\bar{PB}}{\sin(\delta_1)} \quad \frac{\bar{BC}}{\sin(\beta)} = \frac{\bar{PB}}{\sin(\delta_2)} \tag{4}$$

then defining:

$$\gamma_1 = A\hat{B}P \quad \gamma_2 = P\hat{B}C \quad \gamma = \gamma_1 + \gamma_2 \tag{5}$$

we obtain:

$$(\frac{\bar{AB}\sin(\beta)}{\bar{BC}\sin(\alpha)})sin(\delta_1) = sin(\delta_2) = sin(\eta - \delta_1) \quad (6)$$

$$((\frac{\bar{AB}\sin(\beta)}{\bar{BC}\sin(\alpha)}) + \cos(\eta))\sin(\delta_1) = \sin(\eta)\cos(\delta_1) \quad (7)$$

$$\delta_1 = \arctan(\frac{\bar{BC}\sin(\eta)\sin(\alpha)}{\bar{AB}\sin(\beta) + \bar{BC}\cos(\eta)\sin(\alpha)}) \quad (8)$$

and we can compute:

$$\gamma_1 = \pi - \delta_1 - \alpha \quad K = \|\bar{PA}\| = \bar{AB}\frac{\sin(\gamma_1)}{\sin(\alpha)} \quad (9)$$

Knowing θ (the angle between $\bar{BA}$ and the X-axis), we can finally find the position of P:

$$x = K\cos(\delta_1 + \theta) \quad y = K\sin(\delta_1 + \theta) \quad (10)$$

and from the position of P the computation of the orientation is immediate.

For the static localization, at least three beacons are needed: the limits of this algorithm are in the measuring accuracy of the angles and in the beacons position accuracy. When more beacons are seen, a better localization can be made by estimation based on redundant data. As shown in the simulations, this can be done using heuristics. Note that the position of P with respect to the triangle ABC is not important, under the condition that:

$$0 < \alpha < \pi \quad 0 < \beta < \pi \quad (11)$$

Singular configurations occur in the following cases: (i) when P lies on the line between two beacons and (ii) when P, A, B, C lie on the same circle. The former type of singularity is not relevant because in such case the DLPS hardware itself fails (both beacons are obscuring each other). In the latter case, the localization procedure becomes geometrically indefinite, and a preventive check is required.

Dynamic localization is made in an incremental way, updating the odometric position estimated for each new measure. The developed technique is an odometry error correction algorithm. The idea is to reduce intermittently the position and orientation uncertainty of the odometry estimation, which grows with time, as new optical measurements become available. In particular, we define a region around the nominal odometric position, where the robot is supposed to be. The size of this region increases with the distance covered from the last optical measurement and, when a new optical measurement becomes available from another beacon, the system verifies if the measured beam cuts the uncertainty region. If the answer is positive, the new robot position is estimated, resetting the size of the uncertainty region; other-

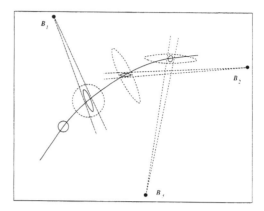

Fig. 5. Scheme of the dynamic localization: for every beacon measured, the uncertainty is reduced in the directon defined by the line that links the beacon to the robot position

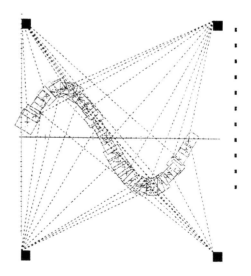

Fig. 6. Simulation of the dynamic localization algorithm: the trajectory is sampled at fixed frequency and, at each step, the uncertainty region, shown as a rectangle, increases in a linear way until a new beacon angle is measured. The position is then corrected, and the uncertainty region is reset to a smaller size.

wise, it means that the odometry error is out of the range and thus it is safer to stop the motion and use the triangulation algorithm to reset the position estimate. At each beacon answer the error is reduced in the direction defined by the line between the beacon and the robot position (see figures 5 and 6). An homogeneous distribution of beacons in the area allows a costant positional error.

4. SIMULATION AND PROTOTYPE TEST

Simulation and tests on the system both on static and dynamic localization have been carried out. The tests done on the real environment confirmed the results of the simulations. Starting from the errors measured on the real prototype (0.3 deg. of uncertainty in the angular measure), the localization accuracy has been tested considering random angular errors with the maximum value of 0.3 deg, and (the worst case) combining the errors on the three beacons. In figure 7 and 8 the simulation results are shown. The grey scale, from white to black express the distance between the real position and the result of the localization algorithm. The test area is about 10x10 mt and the position step of about 3 cm. The random error simulation gives interesting results: the uncertainty circle is fairly visible, and the mean error excluding this critical area is about 2 cm; the worst case obviously gives lower accuracy, 5-7 cm, in the best zones (the light grey lobes).

The real test gives results similar to the random error simulations, with a medium error of 2.5 cm. This simulation gives not only quantitative information on the localization accuracy in the different areas, but is also a tool for beacons positioning in real shop floor industrial applications. As the integration of more than three beacons measures (with angular and positional uncertainty) has no linear implementable solution. We defined an heuristic method for a better positioning accuracy using more than three beacons (see figure 9). On the dynamic localization, we simulated the system imposing random error with a maximum value of 0.3 deg for each beacon, in an area of 10x10 mt and the robot moving at 1.0 mt/sec. The trajectory imposed in the first experiment is a straight line, and in the second experiment a sinusoidal movement with an amplitude of 5 mt (figure 5). The simulation results are 6 cm of error distance in the first case and 8.2 cm of error distance in the second case.

The advantage of this system in comparison with visual pattern recognition techniques is not in the accuracy (that is less) but the unsensitivity to noise, the limited structuration of the environment needed and that we have in any case a constant known approximation of the position, not an

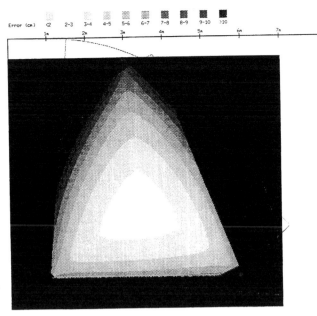

Fig. 7. Localization simulation with 3 beacons, using the localization algorithm based on the Snellius theorem (worst case simulation)

increase of the uncertainty value during the motion.

The evaluation of the results in the testing of the dynamic localization has been more qualitative, for the effective problems in measuring the positions of the robot during the motion. anyway the trajectory defined is accurately executed and the arrival position in the testings is always reached with errors of about 3.0 cm of accuracy (this low error is explained by whe approach to the goal speed, that is always lower than the motion speed, allowing a more accurate localization.

The prototype has been developed at industrial quality level. Both rotating units and beacons are based on a standard microcontroller. The rotating unit uses a class 1b red laser beam. Retransmission by beacons uses infrared non coherent light. In the present, beacons opening range is 60 deg. aperture and 5 mt. distance. or 10 deg. and 15 mt. Information robot-to-beacon and vice versa are exchanged by optical digital modulation. Communication speed is 78 kbit/s robot to beacon, and 2.4 Kbit/s beacon to robot. Information is packed and integrity checked. interleaved with angular data used by localization. A small 4-byte packet is available for general purpose communication among robots and between robots and the host. The throughput is limited. nevertheless is enought to support most of the supervisory activity and cooperation between robots in industrial domain.

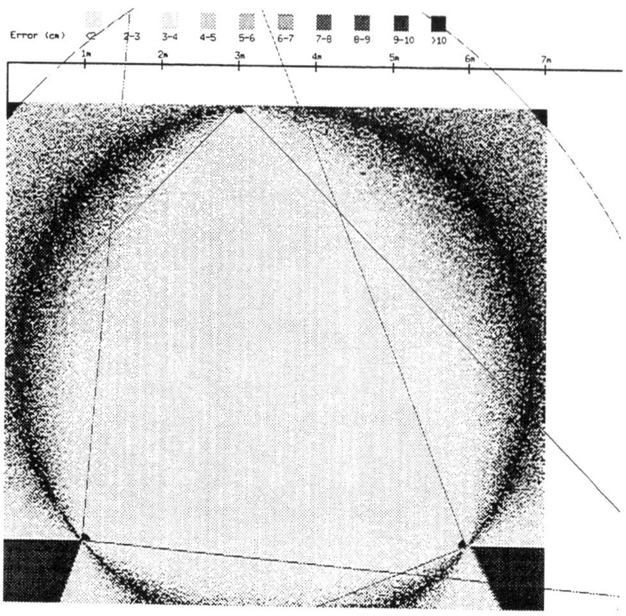

Fig. 8. Localization simulation with 3 beacons, using the localization algorithm based on the Snellius theorem (random error simulation)

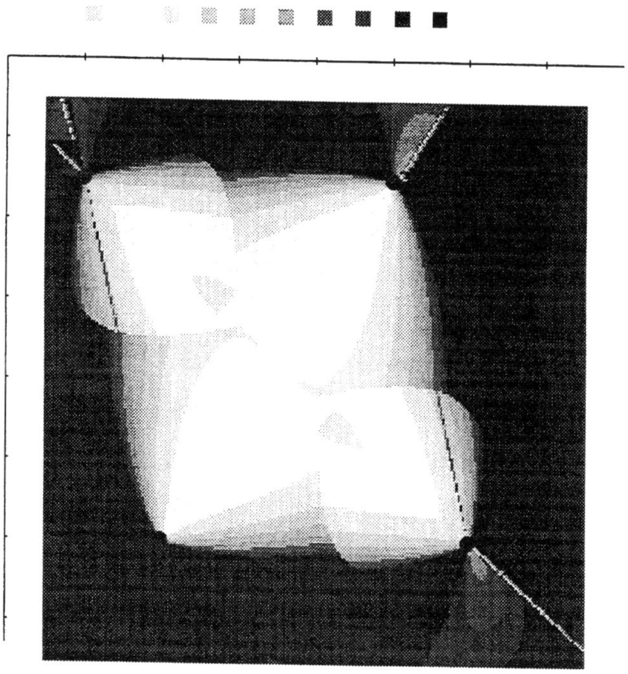

Fig. 9. Localization simulation with 4 beacons, using the localization algorithm heuristics based (worst case simulation)

5. REFERENCES

B.W. Parkinson, S.W. Gilbert (n.d.). Navstar: Global positioning system - ten years later. In: *Proceedings of IEEE.*

C.D.McGillem, T.S.Rappaport (1988). Infra-red location system for navigation of autonomous vehicles. In: *Proc. of ICAR88.* pp. 1236–1238.

Crowley, J.L. (1989). World modeling and position estimation for a mobile robot using ultrasonic ranging. In: *Proceedings of IEEE Conf. on Robotics and Automation.* Vol. 2. pp. 674–680. New York, USA.

Garibotto, G.B. and S. Masciangelo (1992). 3d computer vision for navigation/control of mobile robots. In: *Machine Perception.* Vol. 185. AGARD Lecture Series.

G.L. Miller, E.R. Wagner (1990). An optical rangefinder for autonomous robot cart navigation. In: *Autonomous Robot Vehicles* (Springer Verlag, Ed.).

Masciangelo, S., M. Ilic, A. Camurri and G. Vercelli (1994). A mobile robotic museum guide demonstrating intelligent interaction with the environment and the public. In: *Intl. Symposium on Automotive Technology & Automation.* Aachen, D.

Miller, D.P., A.H. Mishkin, K.E. Lambert, D. Bickler and D.P. Bernard (1989). Autonomous navigation and mobility for a planetary rover. In: *27th Aerospace Sciences Mtg., AIAA 89-0589.*

ABSOLUTE LOCALIZATION OF FAST MOBILE ROBOTS BASED ON AN ANGLE MEASUREMENT TECHNIQUE

U. D. HANEBECK and G. SCHMIDT

Department for Automatic Control Engineering, Technische Universität München, 80290 München, Germany. E-mail: {hnb, gs}@lsr.e-technik.tu-muenchen.de

Abstract. This paper presents an algorithm for absolute localization of mobile robots, which are equipped with an onboard-device making angular measurements on the location of known but undistinguishable landmarks. A simple *linear* solution for the robot position given $N \geq 3$ angle measurements is derived. The associated uncertainties in both landmark positions and angle measurement are modeled as unknown but bounded in amplitude. Experiments with the set theoretic estimator demonstrate its simplicity and effectiveness in real-world applications.

Key Words. Angular measurement; Nonlinear filtering; Recursive Estimation; Set theory; Vehicles

1. INTRODUCTION

This paper introduces an approach for estimating the absolute posture (position x, y; orientation ψ) of a fast mobile robot on a planar surface. The algorithm processes onboard measurements of angular locations of known landmarks. Both, initialization of the robot posture and recursive in-motion posture estimation is considered.

For initialization purposes, a set of angles measured with respect to the robot coordinate system needs to be paired with a subset of the undistinguishable landmarks. In (Wiklund *et al.*, 1988), an enumerative scheme has been reported for pairing the first three angles with landmarks. The remaining angles are used for plausibility tests. Several solutions for calculating the posture given the correct association of measured angles with landmarks have been reported. Sutherland and Thompson (1994), Tsumura *et al.* (1993), and Wiklund *et al.* (1988) consider only triples of landmarks. For the case of more than three landmarks some authors average triple solutions, others use iterative techniques. Betke and Gurvits (1994) supply a closed-form solution for N angles, which does not consider uncertainties. In this paper, a more efficient association algorithm is developed. It discards false measurements, is fast, and is further accelerated by incorporating prior knowledge. Furthermore, it takes advantage of a simple closed-form solution, which consists of a set of $N-1$ *linear* equations for the vehicle position, Sect. 3. An error propagation analysis may be performed, which considers uncertainties in both landmark positions and angle measurement.

If the vehicle velocity is high compared to the angle measurement rate, posture estimates can be sequentially updated by newly incoming bearing measurements. In (Wiklund *et al.*, 1988), a Kalman filtering scheme is introduced for this purpose, which is based on a kinematic vehicle model. White Gaussian random processes are used as uncertainty models; no dead-reckoning information is considered. Nishizawa *et al.* (1995) also use a statistical method to fuse sensor data with dead-reckoning data. In this paper, a set theoretic modeling of the predominantly arising non-random uncertainties is presented. A recursive set theoretic estimator is proposed for the fusion of every measured angle with a posture prediction that is obtained by propagating the previous estimate by means of dead-reckoning data. Sect. 4 suggests an implementation with real-time capabilities. Experiments demonstrate the benefits of the developed localization algorithm in Sect. 5.

2. PROBLEM FORMULATION

Consider a pool of M *undistinguishable* landmarks in a two-dimensional world or map. The positions of the landmarks x_i^{LM}, y_i^{LM}, $i = 0, 1, \ldots, M-1$ in a reference coordinate system are assumed to be known with additive bias errors, which are of course unknown.

$$\begin{aligned}
\hat{x}_i^{\mathrm{LM}} &= \tilde{x}_i^{\mathrm{LM}} + \Delta x_i^{\mathrm{LM}} \\
\hat{y}_i^{\mathrm{LM}} &= \tilde{y}_i^{\mathrm{LM}} + \Delta y_i^{\mathrm{LM}} .
\end{aligned} \tag{1}$$

True values of $*$ will be denoted as $\tilde{*}$, nominal or estimated values as $\hat{*}$. The errors of every landmark are assumed to be confined to an ellipsoidal set given by

$$\begin{pmatrix} \Delta x_i^{\text{LM}} \\ \Delta y_i^{\text{LM}} \end{pmatrix}^T (\mathbf{C}_i^{\text{LM}})^{-1} \begin{pmatrix} \Delta x_i^{\text{LM}} \\ \Delta y_i^{\text{LM}} \end{pmatrix} \leq 1 . \tag{2}$$

Possible correlation of errors for different landmarks is ignored. The robot is capable of determining the angular locations of these landmarks with respect to its coordinate system. Individual landmarks do not necessarily have to be distinguished. The angle measurements are corrupted by additive noise, i.e., $\hat{\alpha} = \tilde{\alpha} + \Delta\alpha$, where $\Delta\alpha$ is assumed to be bounded in amplitude according to $|\Delta\alpha| < \delta_\alpha$. To account for possible occlusion of landmarks in non–convex rooms, partitioning walls are added to the map. The landmarks are ordered in the map so that the robot detects the subset of unoccluded landmarks in that order when scanning counterclockwise.

3. DETERMINING THE INITIAL POSTURE

This section is concerned with (re–) initializing the robot posture $\underline{z} = (x, y, \psi)^T$ when only very little prior knowledge is available. A priori information is specified by confining the posture to an ellipsoidal set $\Omega_{\text{a-priori}}$. M landmarks are available and $3 < N < M$ angles α_i, $i = 0, 1, ..., N-1$ have been measured. The association, i.e., the list of pairings of measured angles to landmarks is initially unknown. Inspired by the interpretation-tree (IT) method in (Drumheller, 1987), the association search is kept from becoming intractable by approaching it in two steps: In the first step, for every measured angle α_i the set of visible landmarks from $\Omega_{\text{a-priori}}$ is determined. In the second step, these visibility constraints are exploited for pruning the IT. Thus, only a small portion of all associations needs to be generated and tested.

Step 1: The projection of $\Omega_{\text{a-priori}}$ onto the x/y-plane is examined at polar grid points $x(r, \theta)$, $y(r, \theta)$ for some r, θ. We define a visibility matrix $\mathcal{V}$ with dimensions N by M. The elements $\mathcal{V}_{ij}$ are boolean variables which are TRUE, if the single measured angle α_i may be caused by landmark j. A visibility test is performed for every grid point $x(r, \theta)$, $y(r, \theta)$. If the landmark j is visible, i.e., when the straight line from the considered grid point $x(r, \theta)$, $y(r, \theta)$ to x_j^{LM}, y_j^{LM} does not intersect any partitioning walls, a hypothetical angle α_{hyp} is calculated. The minimum and maximum angles at $x(r, \theta)$, $y(r, \theta)$ within $\Omega_{\text{a-priori}}$ are denoted as ψ_{LOW}, ψ_{HIGH} respectively. α_i may then be caused by landmark j, if $\alpha_i + \psi_{\text{LOW}} < \alpha_{hyp} < \alpha_i + \psi_{\text{HIGH}}$. If row i of $\mathcal{V}$ does not contain any TRUE value, α_i has been identified as false measurement. Row i is then removed from $\mathcal{V}$ and the number of measurements N is decremented.

Step 2: Only those candidate associations are generated that do not violate the visibility constraints represented by $\mathcal{V}$ and that also follow the ordering assumption. Erroneous measurements are handled efficiently by adopting the "least bad data" constraint proposed in (Grimson and Lozano-Pérez, 1985). For a specific association, a tentative position x_T, y_T is calculated and checked for compatibility with the error bounds, the posture constraint $\Omega_{\text{a-priori}}$, and the requirements for joint visibility of all landmarks involved.

Tentative postures are quickly calculated by use of a closed–form solution. The corresponding set of $N - 1$ *linear* equations for the position is derived next. Define γ_i as the difference between two consecutive angles α_i and α_{i+1}, i.e., $\gamma_i = \alpha_{i+1} - \alpha_i$ or

$$\gamma_i = \text{atan2}(x_{i+1}^{\text{LM}} - x, y_{i+1}^{\text{LM}} - y) \\ - \text{atan2}(x_i^{\text{LM}} - x, y_i^{\text{LM}} - y) . \tag{3}$$

Application of trigonometric identities leads to

$$\tan(\gamma_i) = \frac{\frac{y_{i+1}^{\text{LM}} - y}{x_{i+1}^{\text{LM}} - x} - \frac{y_i^{\text{LM}} - y}{x_i^{\text{LM}} - x}}{1 + \frac{y_{i+1}^{\text{LM}} - y}{x_{i+1}^{\text{LM}} - x} \cdot \frac{y_i^{\text{LM}} - y}{x_i^{\text{LM}} - x}} , \tag{4}$$

which may be rewritten as

$$y_{i+1}^{\text{LM}} y_i^{\text{LM}} + x_{i+1}^{\text{LM}} x_i^{\text{LM}} + \cot(\gamma_i)[x_{i+1}^{\text{LM}} y_i^{\text{LM}} - y_{i+1}^{\text{LM}} x_i^{\text{LM}}]$$
$$= \begin{pmatrix} \cot(\gamma_i)[y_i^{\text{LM}} - y_{i+1}^{\text{LM}}] + x_{i+1}^{\text{LM}} + x_i^{\text{LM}} \\ \cot(\gamma_i)[x_{i+1}^{\text{LM}} - x_i^{\text{LM}}] + y_{i+1}^{\text{LM}} + y_i^{\text{LM}} \end{pmatrix}^T \begin{pmatrix} x \\ y \end{pmatrix}$$
$$- \begin{pmatrix} x & y \end{pmatrix} \begin{pmatrix} x \\ y \end{pmatrix} \tag{5}$$

for $i = 0, 1, ..., N - 1$. Index operations are performed modulo-N, i.e., $i + 1 = 0$ for $i = N - 1$. Subtracting from every equation its follower equation yields a system of $N - 1$ equations that are *linear* in x and y, i.e., $\underline{z} = \mathbf{H}(x, y)^T + \underline{e}$ with $\underline{z} = (z_0, z_1, ..., z_{N-2})^T$, $\mathbf{H} = (\underline{h}_0^T, \underline{h}_1^T, ..., \underline{h}_{N-2}^T)^T$, $\underline{h}_i = (h_i^x, h_i^y)^T$, and error $\underline{e} = (e_0, e_1, ..., e_{N-2})^T$. The corresponding elements are given by

$$z_i = y_{i+1}^{\text{LM}} y_i^{\text{LM}} + x_{i+1}^{\text{LM}} x_i^{\text{LM}} - y_{i+2}^{\text{LM}} y_{i+1}^{\text{LM}} - x_{i+2}^{\text{LM}} x_{i+1}^{\text{LM}}$$
$$+ \cot(\gamma_i)[x_{i+1}^{\text{LM}} y_i^{\text{LM}} - y_{i+1}^{\text{LM}} x_i^{\text{LM}}]$$
$$- \cot(\gamma_{i+1})[x_{i+2}^{\text{LM}} y_{i+1}^{\text{LM}} - y_{i+2}^{\text{LM}} x_{i+1}^{\text{LM}}]$$

$$h_i^x = \cot(\gamma_i)[y_i^{\text{LM}} - y_{i+1}^{\text{LM}}] + x_i^{\text{LM}} \tag{6}$$
$$- \cot(\gamma_{i+1})[y_{i+1}^{\text{LM}} - y_{i+2}^{\text{LM}}] - x_{i+2}^{\text{LM}}$$

$$h_i^y = \cot(\gamma_i)[x_{i+1}^{\text{LM}} - x_i^{\text{LM}}] + y_i^{\text{LM}}$$
$$- \cot(\gamma_{i+1})[x_{i+2}^{\text{LM}} - x_{i+1}^{\text{LM}}] - y_{i+2}^{\text{LM}} .$$

Once x, y are known, ψ can be obtained immediately. To enhance the quality of the solution, an error propagation analysis is performed, which is outside the scope of this paper.

4. IN–MOTION LOCALIZATION

The robot is assumed to be equipped with an odometry which mainly suffers from amplitude–bounded noise sources, that may be strongly correlated or even deterministic. A dead–reckoning system calculates sets of relative posture compatible with the a priori given error bounds. Recursive set theoretic estimation is performed by combining information from an angle measurement at time k with a propagated version of the previous posture estimate at time $k-1$. Ellipsoidal bounding sets (EBS) are used for all operations to achieve real–time capabilities.

4.1. *Propagation*

The result of the fusion process at time $k-1$ is represented by the EBS

$$\Omega_{k-1}^E = \{\underline{z}_{k-1}^E : \tag{7}$$
$$(\underline{z}_{k-1}^E - \underline{\hat{z}}_{k-1}^E)^T \left(\mathbf{C}_{k-1}^E\right)^{-1} (\underline{z}_{k-1}^E - \underline{\hat{z}}_{k-1}^E) \leq 1\}.$$

The dead–reckoning system supplies the set of relative postures with respect to $\underline{z}_{k-1}^E$, henceforth denoted as

$$\Omega_k^\Delta = \{\underline{z}_k^\Delta : \tag{8}$$
$$(\underline{z}_k^\Delta - \underline{\hat{z}}_k^\Delta)^T \left(\mathbf{C}_k^\Delta\right)^{-1} (\underline{z}_k^\Delta - \underline{\hat{z}}_k^\Delta) \leq 1\} .$$

The exact set of absolute postures predicted by the dead–reckoning system is then given by

$$\Omega_k^P = \{\underline{z}_k^P : \underline{z}_k^P = \mathbf{I}\underline{z}_{k-1}^E + \mathbf{B}_k\underline{z}_k^\Delta\} , \tag{9}$$

with $\underline{z}_{k-1}^E \in \Omega_{k-1}^E$, $\underline{z}_k^\Delta \in \Omega_k^\Delta$, $\mathbf{I}$ the identity matrix, and

$$\mathbf{B}_k = \begin{pmatrix} \cos(\psi_{k-1}^E) & -\sin(\psi_{k-1}^E) & 0 \\ \sin(\psi_{k-1}^E) & \cos(\psi_{k-1}^E) & 0 \\ 0 & 0 & 1 \end{pmatrix} . \tag{10}$$

Unfortunately, Ω_k^P is not in general an ellipsoid. Linearizing (9) around the nominal values yields

$$\underline{z}_k^P - \underline{\hat{z}}_k^P \approx \mathbf{J}_k^E(\underline{z}_{k-1}^E - \underline{\hat{z}}_k^E) + \hat{\mathbf{B}}_k(\underline{z}_k^\Delta - \underline{\hat{z}}_k^\Delta) \tag{11}$$

with the Jacobian

$$\mathbf{J}_k^E = \begin{pmatrix} 1 & 0 & -(\hat{y}_k^P - \hat{y}_{k-1}^E) \\ 0 & 1 & (\hat{x}_k^P - \hat{x}_{k-1}^E) \\ 0 & 0 & 1 \end{pmatrix} . \tag{12}$$

Ω_k^P may then be approximated as the EBS for the Minkowski sum of the two ellipsoids in (11)

$$\Omega_k^P \approx \{\underline{z}_k^P : \tag{13}$$
$$(\underline{z}_k^P - \underline{\hat{z}}_k^P) \left(\mathbf{C}_k^P\right)^{-1} (\underline{z}_k^P - \underline{\hat{z}}_k^P)^T \leq 1\} ,$$

with center $\underline{\hat{z}}_k^P$ and $\mathbf{C}_k^P$ given by

$$\underline{\hat{z}}_k^P = \mathbf{I}\underline{\hat{z}}_{k-1}^E + \hat{\mathbf{B}}_k\underline{\hat{z}}_k^\Delta , \quad \mathbf{C}_k^P = \frac{\Xi_k}{1-\kappa} + \frac{\Gamma_k}{\kappa} , \tag{14}$$

with $\Xi_k = \mathbf{J}_k^E\mathbf{C}_{k-1}^E(\mathbf{J}_k^E)^T$, $\Gamma_k = \hat{\mathbf{B}}_k\mathbf{C}_k^\Delta\hat{\mathbf{B}}_k^T$, for $0 < \kappa < 1$ (Schweppe, 1973). κ may be selected such that a measure of the "size" of Ω_k^P is minimized. An appropriate "size" measure is the volume of Ω_k^P, which is proportional to $\sqrt{\det(\mathbf{C}_k^P)}$. Minimizing the volume in three dimensions leads to the problem of determining the unique root of a fourth–order polynomial in $[0, 1]$. The trace measure $\mathrm{tr}(\mathbf{C}_k^P)$, i.e., the sum of the main–diagonal elements of $\mathbf{C}_k^P$, is mathematically most convenient. It leads to the problem of determining the unique root of a second–order polynomial in $[0, 1]$ irrespective of the problem dimensionality. Unfortunately, its meaning is questionable for the problem at hand, where x_k^P, y_k^P and ψ_k^P differ by some orders of magnitude. A more meaningful "size" measure leading to a minimization procedure of intermediate complexity is given by the trace measure of the projection of Ω_k^P onto the xy–plane multiplied by the square of half of the projection of Ω_k^P onto the ψ–axis. This is equivalent to $\mathrm{M}_{\mathbf{C}_k^P\mathbf{C}_k^P}$ with

$$\mathrm{M}_{AB} = \mathrm{tr}\left(\mathrm{proj}_{xy}(\mathbf{A})\right)\mathrm{proj}_\psi(\mathbf{B}) , \tag{15}$$

where $\mathrm{proj}_{SS}(\mathbf{C})$ is defined by eliminating those rows and columns from $\mathbf{C}$ that are not associated with the considered subspace SS. The minimizing κ is given by the unique root of

$$K_3\kappa^3 + K_2\kappa^2 + K_1\kappa + K_0 \tag{16}$$

in $[0, 1]$ where

$$\begin{aligned} K_3 &= 2[\mathrm{M}_{\Xi_k\Xi_k} + \mathrm{M}_{\Gamma_k\Gamma_k} - \mathrm{M}_{\Xi_k\Gamma_k} - \mathrm{M}_{\Gamma_k\Xi_k}] \\ K_2 &= 3[\mathrm{M}_{\Xi_k\Gamma_k} + \mathrm{M}_{\Gamma_k\Xi_k} - 2\mathrm{M}_{\Gamma_k\Gamma_k}] \\ K_1 &= 6\mathrm{M}_{\Gamma_k\Gamma_k} - \mathrm{M}_{\Xi_k\Gamma_k} - \mathrm{M}_{\Gamma_k\Xi_k} \\ K_0 &= -2\mathrm{M}_{\Gamma_k\Gamma_k} . \end{aligned} \tag{17}$$

The proof is straightforward and consists of differentiating the "size" measure with respect to κ and setting the result to zero.

4.2. *Measurement*

The measurement equation for a single α_k at time k and an associated landmark at $x^{\mathrm{LM}}, y^{\mathrm{LM}}$

$$\sin(\alpha_k + \psi_k^M)\{x^{\mathrm{LM}} - x_k^M\}$$
$$= \cos(\alpha_k + \psi_k^M)\{y^{\mathrm{LM}} - y_k^M\} \tag{18}$$

is linearized around the predicted posture $\underline{\hat{z}}_k^P$, the measured angle $\hat{\alpha}_k$, and the nominal landmark position $\hat{x}^{\mathrm{LM}}, \hat{y}^{\mathrm{LM}}$. This yields $r_k = \underline{D}_k^T\underline{z}_k^M + e_k$ with

$$r_k = c_k^1 \hat{x}^{\mathrm{LM}} - c_k^2 \hat{y}^{\mathrm{LM}} - c_k^3 \hat{\psi}_k^P$$
$$\underline{D}_k = (c_k^1, -c_k^2, -c_k^3)^T \tag{19}$$
$$e_k = (-c_k^1, c_k^2)(\Delta x^{\mathrm{LM}}, \Delta y^{\mathrm{LM}})^T - c_k^3 \Delta \alpha .$$

The constants are defined as $c_k^1 = \sin(\hat{\alpha} + \hat{\psi}_k^P)$, $c_k^2 = \cos(\hat{\alpha} + \hat{\psi}_k^P)$, and $c_k^3 = c_k^2\{\hat{x}^{\mathrm{LM}} - \hat{x}_k^P\} + c_k^1\{\hat{y}^{\mathrm{LM}} - \hat{y}_k^P\}$. e_k can thus be bounded, i.e., $|e_k| < e_k^{max}$ with

$$e_k^{max} = \sqrt{\begin{pmatrix} -c_k^1 \\ c_k^2 \end{pmatrix}^T \mathbf{C}^{\mathrm{LM}} \begin{pmatrix} -c_k^1 \\ c_k^2 \end{pmatrix}} + |c_k^3|\delta_\alpha \tag{20}$$

Two equivalent representations of the measurement set will be given in the following. The first consists of two boundary planes given in normalized Hesse form as

$$\underline{N}_k^T z_k^M + t_k^1 = 0, \ \underline{N}_k^T z_k^M + t_k^2 = 0 \tag{21}$$

with

$$\underline{N}_k = -\frac{\underline{D}_k}{\|\underline{D}_k\|}, \ t_k^1 = \frac{r + e_{max}}{\|\underline{D}_k\|}, \ t_k^2 = \frac{r - e_{max}}{\|\underline{D}_k\|} \tag{22}$$

The second representation is

$$\Omega_k^M = \{z_k^M : |\underline{A}_k^T z_k^M - b_k| \le 1\} . \tag{23}$$

They may be converted into each other via $\underline{A}_k = 2\underline{N}_k/(t_k^2 - t_k^1)$, $b_k = -(t_k^2 + t_k^1)/(t_k^2 - t_k^1)$.

4.3. Robust Combination of Information

Conceptionally, fusion just consists of calculating the intersection of the sets Ω_k^P, Ω_k^M. Ω_k^P is the ellipsoidal set of predicted postures, Ω_k^M is the measurement strip. The intersection of the two sets Ω_k^P, Ω_k^M is not in general again an ellipsoid. To arrive at a recursive scheme, an ellipsoid circumscribing the intersection is required. A bounding ellipsoid is given by (Sabater and Thomas, 1991)

$$\Omega_k^E = \{z_k^E : (z_k^E - \hat{z}_k^E)(\mathbf{C}_k^E)^{-1}(z_k^E - \hat{z}_k^E)^T\}$$
$$\mathbf{C}_k^E = K_k \mathbf{C}_k^0$$
$$\mathbf{C}_k^0 = \mathbf{C}_k^P - \lambda \frac{\mathbf{C}_k^P \underline{A}_k \underline{A}_k^T \mathbf{C}_k^P}{1 + \lambda G_k}$$
$$\hat{z}_k^E = \hat{z}_k^P + \lambda \mathbf{C}_k^0 \underline{A} \epsilon_k \tag{24}$$
$$\epsilon_k = b_k - \underline{A}_k^T \hat{z}_k^P$$
$$G_k = \underline{A}_k^T \mathbf{C}_k^P \underline{A}_k$$
$$K_k = 1 + \lambda - \lambda \epsilon_k^2/(1 + \lambda G_k)$$

for all $\lambda \ge 0$. This set has the interesting property that it both contains the intersection of the measurement and the prediction set and is itself contained in their union, i.e.,

$$(\Omega_k^M \cap \Omega_k^P) \subset \Omega_k^E \subset (\Omega_k^M \cup \Omega_k^P) . \tag{25}$$

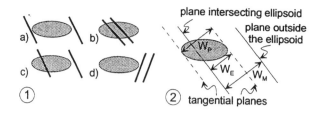

Fig. 1. 1: Configurations for strip and ellipsoid (2D).
2: Definitions for assessing consistency.

The non-linear fusion comprises 4 cases:

- **No uncertainty reduction:** The actual measurement is of no help in reducing the uncertainty, Fig 1.1 a).
- **Consistency:** Both planes defining the measurement set intersect the ellipsoidal prediction set, Fig. 1.1 b).
- **Partial consistency:** Only one plane intersects the prediction set, Fig. 1.1 c).
- **Inconsistency:** Prediction set Ω_k^P and measurement set Ω_k^M do not share a common point, Fig. 1.1 d).

For the case of (partial) consistency of ellipsoid and strip, the *volume* of the bounding ellipsoid in (24) may be minimized by selecting the weight λ as the most positive root of the quadratic equation given by (Sabater and Thomas, 1991)

$$(N-1)G_k^2\lambda^2 + (2N - 1 + \epsilon_k^2 - G_k)G_k\lambda$$
$$+N(1 - \epsilon_k^2) - G_k = 0 , \tag{26}$$

where N is the dimension, here $N = 3$. This result is now generalized to obtain an EBS with a *minimum volume projection* onto an arbitrary subspace. λ_{OPT} is the positive real root of

$$\lambda^3(G_k - H_k)G_k^2 L + \lambda^2\{L(3G_k - 2H_k) - H_k\}G_k$$
$$+ \lambda\{[3G_k - H_k + \epsilon_k^2(H_k - G_k)]L \tag{27}$$
$$- G_kH_k - H_k + \epsilon_k^2H_k\} + L(1 - \epsilon_k^2) - H_k ,$$

with L the subspace dimension and

$$H_k = \underline{A}_k^T(\bar{\mathbf{C}}_k^P)^T[\mathrm{proj}(\mathbf{C}_k^P)]^{-1}\bar{\mathbf{C}}_k^P\underline{A}_k^T . \tag{28}$$

$\bar{\mathbf{C}}_k^P$ is obtained from $\mathbf{C}_k^P$ by eliminating the rows not associated with the considered subspace. The proof is patterned after the one in (Deller Jr and Luk, 1989) and is sketched in the appendix. Application of this concept to the localization problem leads to a tailor-made bounding operation. The inherently high precision of the orientation estimate ψ^E compared to the position estimate x^E, y^E is considered by minimizing the projection of the EBS onto the x, y subspace. The resulting EBS is more conservative in ψ^E, but tight for the more critical position estimate x^E, y^E.

Using the optimal EBS is successful as long as the model is exact. However, modeling errors may

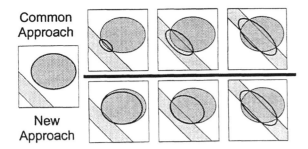

Fig. 2. EBSs for the intersection of ellipsoid and strip. Top: Common scheme. Bottom: Extension employing consistency measures.

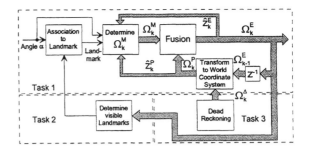

Fig. 3. Scheme of set theoretic recursive estimator.

lead to an unreasonably small estimation set Ω_k^E. Enhanced robustness is achieved by imposing a higher priority on the set of predicted states Ω_k^P since it contains the information obtained from all the past measurements. This priority should depend on the degree of consistency of the two sets Ω_k^P and Ω_k^M. Roughly speaking, the idea is to select the set Ω_k^E such that it exhibits a growing tendency towards the prediction set Ω_k^P with falling degree of consistency of the sets Ω_k^P and Ω_k^M. Referring to Fig. 1.2, a reasonable consistency measure is given by the intersection width W_E divided by the geometric mean of the strip width W_M and the ellipsoid width W_P

$$\mathrm{CM}(\Omega_k^P, \Omega_k^M) = \frac{W_E}{\sqrt{W_P W_M}}, \ 0 \le \mathrm{CM} \le 1. \quad (29)$$

λ in (24) is selected from $[0, \lambda_{\mathrm{OPT}}]$ as an appropriate function of the consistency measure. A shifted logistic function

$$\lambda = \lambda_{\mathrm{OPT}}/[1 + \exp(-S(\mathrm{CM} - M))], \quad (30)$$

is used with S, M chosen as $S = 10$, $M = 0.5$. The influence of this extension on the fusion result is demonstrated in Fig. 2 by comparing it with the common approach for four cases. For the common approach, the volume of the resulting EBS Ω_k^E experiences large changes when the measurement set just changes slightly. Single (unmodeled) measurement outliers lead to an extremely small EBS. On the other hand, the new approach calculates Ω_k^E by modifying Ω_k^P depending on its consistency with Ω_k^M. Thus, single erroneous measurements have a reduced impact on the fusion result.

A schematic overview of the proposed scheme for localization during fast motion based on angle measurements is depicted in Fig. 3. The feedback of $\hat{z}_k^E$ to the process for determination of Ω_k^M deserves some attention. It may replace $\hat{z}_k^P$ to iteratively refine the linearization of (18). The scheme may easily be parallelized into three tasks: The fusion loop, the determination of landmarks not occluded by partitioning walls, and dead-reckoning.

5. EXPERIMENTAL VALIDATION

The effectiveness of the new approach is demonstrated by navigating a fast (2 m/sec) omnidirectional service robot (Hanebeck and Schmidt, 1994) through an office environment. An onboard laser-based goniometer takes angular measurements on retro-reflecting tape strips attached to the walls as artificial landmarks. 20 horizontal 360°-scans per second are performed; absolute accuracy is about 0.02°. Thus, the predominant uncertainties result from inaccurate knowledge about the positions of the fixed landmarks. A map contains nominal positions of 22 identical landmarks and partitioning walls, Fig. 4. The full scale robot is equipped with three independently steerable drive wheels. Dead-reckoning based on these wheels suffers from error sources like imperfect wheel coordination and uncertain wheel/floor contact points. The usual white Gaussian noise model is not appropriate here. The amplitude-bounded nature of the correlated error sources suggests the proposed set theoretic treatment. Once initialized with the algorithm introduced in Sec. 3, the robot performs 10 cycles of a predefined course, Fig. 4. The localization estimate based on the fusion of goniometer data and dead-reckoning is compared with data from dead-reckoning only. The highly correlated nature of the accumulating dead-reckoning errors is obvious. On the other hand, the vehicle is kept very accurately on track with the localization estimate. Absolute accuracy has found to be about ±2 cm and ±0.5°.

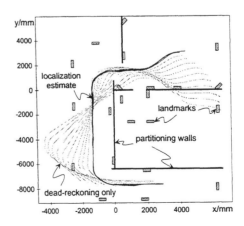

Fig. 4. Experimental in-motion localization.

6. CONCLUSION

Set theoretic concepts have been applied to posture estimation of fast-moving mobile robots which perform angular measurements on the location of known but undistinguishable landmarks. In this context, four main results are presented:

1) A simple *closed-form* solution for the robot position given angular locations of N known landmarks is derived, which consists of $N-1$ equations *linear* in the position.

2) Consistency measures for assessing the consistency of candidate posture sets are introduced to design set theoretic estimators which are robust to modeling errors.

3) The common minimum volume EBS algorithms for the intersection of an ellipsoid and a strip are extended to obtain the EBS with *minimum volume projection* onto an arbitrary subspace. Application to the localization problem allows consideration of the inherently higher precision in orientation ψ by selecting the EBS with *minimum volume projection* onto the xy-plane. This EBS bounds the exact estimation set more tightly in x, y and is more conservative in ψ.

4) A scheme for approximate propagation of an ellipsoidal posture set with set valued data from an uncertain dead-reckoning system is given, which just involves finding the roots of a third-order polynomial.

The effectiveness of the proposed set theoretic estimator has been demonstrated by experiments with a fast omnidirectional service robot. The full scale robot is equipped with a laser-based goniometer which makes angular measurements on the location of tape strips attached to the wall as artificial landmarks. Navigation in an office environment revealed a high absolute accuracy of about ± 2 cm and $\pm 0.5^0$.

7. REFERENCES

Betke, M. and L. Gurvits (1994). Mobile Robot Localization Using Landmarks. In: *Proc. of the 1994 IEEE/RSJ/GI Int. Conf. on Intelligent Robots and Systems, Munich, Germany.* pp. 135–142.

Deller Jr, J. R. and T. C. Luk (1989). Linear Prediction Analysis of Speech Based on Set–Membership Theory. *Computer Speech and Language* **3**, 301–327.

Drumheller, M. (1987). Mobile Robot Localization Using Sonar. *IEEE Trans. on PAMI* **9**(2), 325–332.

Grimson, W. E. L. and T. Lozano-Pérez (1985). Recognition and Localization of Overlapping Parts From Sparse Data in Two and Three Dimensions. In: *Proc. of the 1985 IEEE Int. Conf. on Robotics and Automation, St. Louis, MO.* pp. 61–66.

Hanebeck, U. D. and G. Schmidt (1994). A New High Performance Multisonar System for Fast Mobile Robots. In: *Proc. of the 1994 IEEE/RSJ/GI Int. Conf. on Intelligent Robots and Systems, Munich, Germany.* pp. 1853–1860.

Nishizawa, T., A. Ohya and S. Yuta (1995). An Implementation of On-board Position Estimation for a Mobile Robot. In: *Proc. of the 1995 IEEE Int. Conf. on Robotics and Automation, Nagoya, Japan.* pp. 395–400.

Sabater, A. and F. Thomas (1991). Set Membership Approach to the Propagation of Uncertain Geometric Information. In: *Proc. of the 1991 IEEE Int. Conf. on Robotics and Automation, Sacramento, CA.* pp. 2718–2723.

Schweppe, F. C. (1973). *Uncertain Dynamic Systems.* Prentice–Hall.

Sutherland, K. T. and W. B. Thompson (1994). Localizing in Unstructured Environments: Dealing with the Errors. *IEEE Trans. on Robotics and Automation* **10**(6), 740–754.

Tsumura, T., H. Okubo and N. Komatsu (1993). A 3-D Position and Attitude Measurement System Using Laser Scanners and Corner Cubes. In: *Proc. of the 1993 IEEE/RSJ Int. Conf. on Intelligent Robots and Systems, Yokohama, Japan.* pp. 604–611.

Wiklund, U., U. Andersson and K. Hyyppä (1988). AGV Navigation by Angle Measurements. In: *Proc. of the 6th Int. Conf. on Automated Guided Vehicle Systems, Brussels, Belgium.* pp. 199–212.

8. APPENDIX

Proof of (27): The projection of $\mathbf{C}_k^E$ onto a certain subspace is denoted as $\text{proj}(\mathbf{C}_k^E)$ and may be written as

$$K_k \, \text{proj}(\mathbf{C}_k^P) - \lambda K_k \frac{\text{proj}(\mathbf{C}_k^P A_k A_k^T \mathbf{C}_k^P)}{1 + \lambda G_k} \ . \quad (31)$$

The volume of $\text{proj}(\mathbf{C}_k^E)$ is proportional to

$$\det\left(K_k \mathbf{I} - \lambda K_k \frac{[\text{proj}(\mathbf{C}_k^P)]^{-1} \bar{\mathbf{C}}_k^P A_k A_k^T \bar{\mathbf{C}}_k^P}{1 + \lambda G_k}\right) \quad (32)$$

where $\bar{\mathbf{C}}_k^P$ is defined as $\mathbf{C}_k^P$ with the rows not associated with the subspace eliminated. Applying the matrix identity (Deller Jr and Luk, 1989)

$$\det(c\mathbf{I} + \underline{y}\underline{z}^T) = c^{L-1}(c + \underline{y}^T \underline{z}) \ , \quad (33)$$

where L is the subspace dimension, (32) becomes

$$K_k^L \frac{1 + \lambda(G_k - H_k)}{1 + \lambda G_k} \ , \quad (34)$$

with H_k from (28). Differentiating with respect to λ and setting the result to zero yields (27).

VEHICLE POSITION ESTIMATION THROUGH VISUAL
TRACKING OF FEATURES IN A STREAM OF IMAGES[1]

J. Ferruz and A. Ollero.

Escuela Superior de Ingenieros. Universidad de Sevilla.
Departamento de Ingeniería de Sistemas y Automática.
Avenida Reina Mercedes, 41012 Sevilla (Spain).
Fax:+34-5-4556849. Email: ferruz@esi.us.es and aollero@esi.us.es

Abstract. This paper deals with vehicle´s position estimation using an on board vision system. Particularly, a sequence of images provided by a conventional video camera is used to estimate the motion. The method is based on visual tracking of simple features (windows) in the stream of images. Then, the new position is estimated from the computed motion of the tracked features. The features tracking algorithm includes, as significant characteristics, the search of a starting point to reduce the number of candidates, the computation of correlations using grey levels, and the solving of multiple matching by choosing first best fitting pairs. Procedures to eliminate spurious features due to noise, and to avoid erroneous position estimation due to poor feature information have been implemented. The method have been applied in outdoor environments. Particularly, a stream of images for position estimation of an autonomous vehicle inside an electrical workstation has been considered.

Keywords: Position estimation of vehicles, visual tracking, mobile robots.

1. INTRODUCTION

Position estimation is a basic function for autonomous vehicles and mobile robots. This function is necessary to accomplish several tasks including path planning, trajectory generation, and obstacle avoidance.

Dead reckoning is the simplest approach in estimating the vehicle position and orientation. In this case only the counting of the wheels revolutions by means of optical encoders is needed. The procedure is simple and inexpensive but the uncertainty in the estimation grows very fast.

The Inertial Navigation System provides position estimation, but also leads to large inaccuracies over time.

One approach for position estimation that can be used with precision enough for several applications is to structure the environment with beacons, such as lights, infrared emitters, radio beacons, and so on. If a sufficient number of beacons is "visible" during navigation, triangulation can be applied for position estimation. However, structuring the environment with beacons may be costly and create maintenance problems. Furthermore, this procedure lacks generality since it requires alterations in the environment. This lack of generality is also common if natural landmarks are used instead of man made beacons.

An alternate approach is the matching of environment data in the current vehicle's position to the corresponding data in a previously known model. Particularly, active sensors, such as ultrasonic or laser sensors can be used to obtain these data. In (Gonzalez, Stenz and Ollero, 1992) a scanner laser range finder is used for mobile robot precise position estimation.

However, active sensors have some drawbacks. Thus, the number of sensors required to "see" all the environment may be high. Alternatively, a scanning mechanism can be used. However, these mechanism should be very precise. Then the cost could be high and also have significant maintenance problems. Furthermore, active sensors give problems when several sensors are transmitting and receiving signals (several sensors in the same or in different vehicle).

Thus, in many applications, passive sensors such as cameras, are preferred. However, the problem is the computational complexity of the procedures to extract and match image features when the environment is not highly structured.

2. DESCRIPTION

The position estimation method proposed in this paper is based on features extracted from a sequence of images. More precisely, the following stages in the estimation process are considered:

- Feature detection and selection.

- Feature tracking.

[1]This research has been partially supported by the CICYT project TAP93-0581 and the Dirección General de Industria de la Junta de Andalucía.

- Position estimation.

In the following, these stages will be summarized.

Feature detection and selection.

The goal is to identify the motion of the camera that accounts for the changes detected between images, not to recover the shape of a surface. Then a limited number of highly reliable features is selected and tracked through the image stream.

The features selected are the grey levels found in small windows within the first image of a series. This approach has been chosen because it proves to be reliable in a wide range of images and requires only local processing. Environments are more easily recognized than single point corners, and they seem less costly to detect than other structures such as straight lines.

The algorithm used to select the features is based on Tomasi (91). Given the grey levels (I, J) found in a pair of images, the goal is to minimize the error:

$$\varepsilon = \int_w (I(x - \delta) - J(x))^2 dA \qquad (1)$$

where δ is the displacement from the original position in the first image and the surface integral is extended to the whole window. If a linear approximation is used for the grey levels, the value of δ that minimizes (1) is given in closed form by the 2x2 system:

$$G\delta = e \qquad (2)$$

where

$$G = \int_w gg^T dA \qquad (3)$$

$$e = \int_w (I - J) g dA \qquad (4)$$

and g is the computed intensity gradient at a given point of the original image.

The displacement can be obtained by iteratively solving the linear system (2). The features are chosen so that the matrix G assures a well-conditioned solution; in practice, a minimum value for the smaller eigenvalue is imposed.

Feature tracking.

The iterative linear algorithm used by Tomasi works well if δ is small when compared with the size of the window. If the new position of a window is far from the preceding, to the point that little or no overlap exists, this simple approach cannot be used.

In real problems such as the tracking of features in video images obtained from a mobile robot, the method has to be modified to deal with larger displacement vectors. Otherwise, the maximum speed of the platform would be very low. The tracking algorithm has been modified by including the search of a suitable initial start point for the iterative method. The basic steps are the following:

- Define a limited search environment, so that the number of candidates is reduced.

- Feature correlation: A set of candidate windows is selected in the second image with the same criteria used for the first. The features inside the search environment of every tracked window are compared with it. If the initial error is below a threshold, the would-be matching is considered in the next step; otherwise it is discarded.

- Grey-level correlation: The iterative algorithm is applied from the start points given by the set of valid candidates for any given window in the first image. Multiple matchings are solved by choosing the best-fitting pairs first. Incompatible matchings are jointly considered, so that only the best of each class is accepted.

- Local support evaluation: Once the best matchings are chosen by means of the minimum error criterion, a final test is performed. For each window tracked, a cluster of neighbors is searched for that keeps the same shape as in the preceding frame, up to a scale factor. If a large enough cluster is found, every candidate in it is chosen for the corresponding window; if not, the matching of the initial candidate is discarded.

The last two steps may be iterated until a decision is taken about every possible matching.

Position estimation.

Once the correspondence of points has been established in at least two images, it can be used to estimate motion. The relation between the unknown coordinates of features relative to the camera in the second and first images is:

$$x_2 = Rx_1 + T \qquad (5)$$

where R is the rotation matrix and T the translation vector. If a sequence of monocular images is used, only the direction of T can be determined, so that the number of computable motion parameters is five. Once T and R are determined for a given pair of images, the 3D positions of the features can be found. For monocular images both the positions of the points and the speed of motion can be computed up to a scale factor.

The following algorithm is proposed:

1: Capture first image, which defines the start position.

2: Select windows to be tracked over the next images.

3: Repeat until end condition:

 3.1: Repeat while the number of windows surviving from the last position estimation is greater than a certain threshold:

- Capture image.

- Track the current set of windows.

- Choose new windows to be tracked in order to replace the lost ones.

84

3.2: Estimate motion between the last known position only with the surviving windows.

3.3: Estimate positions of old and new windows.

The steps 3.2 and 3.3 require a method to recover motion parameters from a pair of images. Several techniques can be applied. There is a well-known and relatively simple closed-form method that uses the epipolar constraint (see Weng et al. (1989) for a full description). The so-called 8-point algorithm allows to solve the non-linear problem by calculating a set of intermediate parameters (the "essential matrix") linearly dependent on the image data. The procedure is simple and fast, but very sensible to noisy data. Several methods have been proposed to improve its robustness (Weng et al. (1989 and 1993), Zhang et al. (1994)). In its simplest form, the algorithm uses the constraint:

$$(x_2)^t (T \times (Rx_1)) = 0 \qquad (6)$$

which means that the vectors x_2, Rx_1 and T are coplanar. (6) can be shown to be equivalent to

$$(x_2)^t Ex_1 = 0 \qquad (7)$$

This equation defines a linear system in the elements of E, the essential matrix. From E both R and the direction of T can be obtained.

An alternative approach (see Weng et al. (1993)) is to find the motion and structure parameters that minimize the sum of squared errors on the image plane between the observed and computed projections:

$$\varepsilon = \sum_{i=1}^{n} (\|u_i - u_{ci}(m, x)\|^2 + \|v_i - v_{ci}(m, x)\|^2) \qquad (8)$$

where n is the number of windows, u_i, u_{ci} are the observed and computed projections for the i feature in the first image and v_i, v_{ci} represent the same for the second image. m and x are the parameters whose value should minimize ε: $3n$ coordinates for the tracked features and five additional parameters to define position (undetermined by a scale factor) and orientation of the camera in the second image. To minimize (8) a partial standard minimization on the motion parameters is combined with an approximate closed-form solution for feature coordinates. The closed form solution is used to find a good starting point.

A method related to squared error minimization is an iterated extension for non-linear systems of Kalman filtering. This kind of parameter estimation has been used to estimate motion, and its results has been analysed (Weng et al. 1992, Weng et al. 1993). It has been found that sequential approaches lead to poor results due to divergence; a batch processing is needed where a sufficiently large set of data is jointly considered.

For applications such as mobile robot guidance is necessary to keep an updated estimation of uncertainty, and to consider additional data such as dead-reckoning to stabilize motion estimation. This makes Kalman filtering a potentially useful technique for vehicle guidance applications where additional data are available to improve results.

3. IMPLEMENTATION

The algorithms presented in section 2 for feature detection, tracking and position estimation have been developed and implemented on a Sun workstation integrated in the mobile robot controller. Currently the system uses the simple and fast 8-point algorithm for the motion parameter estimation, but Kalman filtering techniques will be probably used in the future.

A more powerful implementation has been designed. This implementation is based on a set of 320C40 DSP processors interconnected through their built-in communication channels to allow near real-time parallel image processing and motion estimation. In this implementation a stream of image frames is digitized and directly sent through a communication channel, so that processing can start before the whole image has been received. A description of a similar system can be found in Amidi (1993).

4. EXPERIMENTAL RESULTS

The results of tracking and motion estimation for three pairs of images are included. 10x10 pixel windows are tracked; the recovered translation and rotation values are interpreted as in (5). The reference system is fixed to the camera: Origin on the focal point, Z axis on the optical axis, Y axis pointing to the right side of the image and X axis pointing to the upper end of the image. In all three cases the simple 8-point algorithm was used for motion recovery.

The two first experiments (figures 1 to 4) are outdoor scenes. Both were recorded with a low resolution home-video camera. Two frames were digitized. The 256x256 images were generated by discarding one of the fields that make up a frame and half of the original 512 pixels of each line.

The "car" sequence (figures 1 and 2) was digitized while the camera was kept motionless in the first and last position. The motion consisted of a forward translation along the Z axis without rotation of the camera. Both images are very noisy; a low-pass filtering was necessary before starting eigenvalue calculation to avoid the selection of useless windows. The results of the tracking are good; there are no false matchings and most of the lost windows were noisy and unreliable. The calculated translation and rotation are fairly consistent with the real motion:

Translation direction vector:

x: 0.002 y: -0.018 z: -0.999

Camera rotation:

yaw: 0.723 pitch: -0.519 roll: 0.775(degrees)

The second experiment (figures 3 and 4) has been performed in the framework of a project on the inspection of electrical substations by using mobile robots. It is intended to provide the robot with a position estimation algorithm for autonomous navigation in the substation.

In this case the video sequence was recorded while the camera was moving, and the motion was known to be roughly forward, with small rotation of the hand-held camera. The quality of this pair of images is not good either; in addition the near-periodic structures in the upper side make easy to get a bad match, and occlussions lead to great local changes given the significant displacement between both images. Only a fraction of the selected windows can be found; anyway most matches are good and there is still a set of points sufficient to compute motion. The absence of local support eliminates most false matches. The results are consistent with the real motion:

Translation direction vector:

x: -0.093 y: -0.17 z: -0.981

Camera rotation:

yaw: -2.34 pitch: 1.26 roll: -2.365(degrees)

The last experiment (figures 5 and 6) uses an indoor sequence. The images were directly digitized from a surveillance camera. The motion was controlled and consisted of a forward translation parallel to Z axis without rotation. The tracking is good between both images for the surviving windows; no false matches are generated. Many of the lost windows have disappeared off the field of view. Others have changed too much because they are too close to the camera. The recovered motion is close to the real one:

Translation direction vector:

x: 0.074 y: 0,043 z: -0.996

Camera rotation:

yaw: -0.269 pitch: -0.009 roll: -1.088(degrees)

5. CONCLUSIONS

In this paper a method for position estimation of autonomous vehicles and mobile robots has been presented. The method is based on visual tracking of features in a stream of images given by a conventional video camera. The method first applies a tracking algorithm to compute a set of features in a new image and then estimates the position from the computed features.

The feature-tracking algorithm improves previous methods by including the search of a suitable initial start point for the iterative method so that the number of candidates is reduced. Then applies feature correlation and finally uses grey level correlation. Multiple matchings are solved by choosing the best-fitting pairs first and considering the local support for each possibility.

An algorithm is proposed for position estimation from the tracked windows. This algorithm works on sets of data taken from positions placed as far as possible from each other to minimize the influence of noise.

The method has been applied both in the laboratory and in outdoor environments as encountered in a project on inspection of electrical substations by means of a mobile robot.

The experimental results are very promising. The current implementation on a Sparc workstation integrated in the mobile robot controller will be greatly improved by means of a future parallel implementation with a set of 320C40 DSP processors

and a digitizer sending directly the images so that processing can start before the whole image has been received.

6. ACKNOLEWDGEMENTS

The authors want to acknowledge the assistance for the electrical substation experiment from Pedro Zarco, Antonio Lopez and other personnel from the "Compañía Sevillana de Electricidad".

7. REFERENCES

Amidi, O., Mesaki, Y. and Kanade, T. (1993), "Research on an Autonomous Vision-Guided Helicopter", Robotics Institute, Carnegie Mellon University, Pittsburgh, PA. USA.

Evans, R. (1992), "3D Computer Vision Techniques for Object Following and Obstacle Avoidance", Roke Manor Research Limited, Roke Manor Romsey, Hampshire, UK.

Gonzalez, J., Stentz, A. and Ollero, A. (1992), "Mobile Robot Iconic Position Estimator using a Radial Laser Scanner", to be published in *Journal of Intelligent and Robotic Systems*, 1995.

Huang, T.S. and Faugeras, O. D. (1989), "Some Properties of the E Matrix in Two-View Motion Estimation", PAMI NO. 12, pp. 1310-1312.

Lucas, B. D. and Kanade, T. (1981), "An Iterative Image Registration Technique with an Application to Stereo Vision", in *Proceedings of the 7th International Joint Conference of Artificial Intelligence.*

Rehg and Witkin (1991), "Visual Tracking with Deformation Models". In *Proceedings of the IEEE International Conference on Robotics and Automation*, pp. 844-850, Sacramento, CA, April 1991.

Tomasi, C. (1991), "Shape and Motion From Image Streams: A Factorization Method", Ph.D. Thesis, Carnegie Mellon University.

Weng, J., Huang, T.S., and Ahuja N. (1989), "Motion and Structure from Two Perspective Views: Algorithms, Error Analysis, and Error Estimation", PAMI NO. 5, pp. 451-475.

Weng, J., Huang, T.S. and Ahuja, N. (1992), "Motion and Structure from Line Correspondences: Closed-Form Solution, Uniqueness, and Optimization", PAMI NO. 3, pp. 362-382.

Weng, J., Cohen, P. and Rebibo N. (1992), "Motion and Structure Estimation from Stereo Image Sequences", PAMI NO. 3, pp. 362-382.

Weng, J., Ahuja, N. and Huang T. S. (1993), "Optimal Motion and Structure Estimation", PAMI NO. 9, pp. 864-884.

Zhang, Z., Deriche, R., Faugeras, O., and Luong, Q.T. (1994), "A Robust Technique for Matching Two Uncalibrated Images Through the Recovery of the Unknown Epipolar Geometry", Rapport de recherche, INRIA.

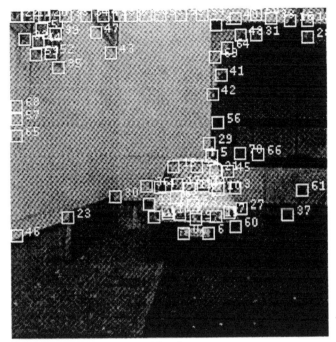

Figure 1. Car. First image.

Figure 3. Electrical substation. First image.

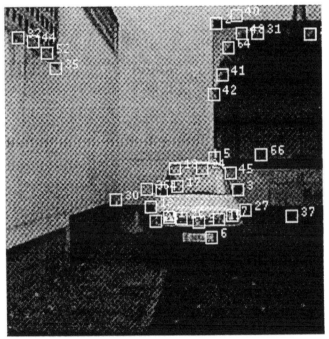

Figure 2. Car. Second image.

Figure 4. Electrical Substation. Second image.

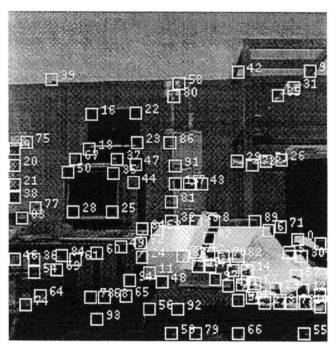

Figure 5. Lab test. First image.

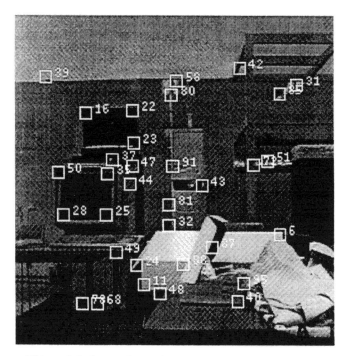

Figure 6. Lab test. Second image.

PAVCAS: A CAR DRIVING SIMULATOR FOR HUMAN BEHAVIOR ANALYSIS

J. CLOT, T.PEBAYLE, T.SENTENAC

LAAS-CNRS
Laboratoire d'Analyse et d'Architecture des Systemes
7 avenue du colonel Roche
31077 Toulouse
Fax: 61 33 62 08
Tel: 61 33 62 00
e.mail: laas-contact@laas.fr

Abstract: The driving simulator PAVCAS is designed for the study of human driving behaviour. It takes place in the general PAVCAS research program. Its aim is to develop a specific module ,embarked in a vehicle, which will warn a sick or tired driver if he has an abnormal or dangerous driving behavior. To achieve this goal we have built a specific simulator including a motion platform. The purpose of this paper is to present and discuss the technological solution adopted, its software and hardware implementation and its assessment.

INTRODUCTION

The context

Historically, simulation started with the ground based flight trainers. The simulator was obviously less expensive and much safer to operate compared to training in real environment.

Simulators are widely used in aircraft, transport, space, not only for training but also for studying and designing the simulated vehicle itself. In this respect, simulation is encouraged by the constant increase of computers performances.

Nowadays, driving simulation is realistic enought to suggest that a simulator could be a tool for the study of human behavior when experiments in real environment would be impracticable or dangerous.

PAVCAS research program

The PAVCAS research program (study of human vigilance in a driving task, using a driving simulator) is developed in two laboratories of the French National organization for Scientific Research: LAAS and LPPE (Laboratoire de Physiologie et de Psycologie Environnementale).

The aim of this program is to develop a specific module to analyze the driver behavior. This module is the first step to design a 'black box', embarked in a vehicle, which will tell a sick or tired driver if he has an abnormal or dangerous driving behavior.

Two analysis will help us to design this module:
- The first analysis will rule out any kind of sensors on the driver's body. For example, we shall focus our interest to the steering wheel movements, lateral position, reaction time of the driver,... The analysis of these data will associate various technics such as neural computation.
-The second analysis is a psychological and physiologic study of human driver at LPPE laboratory.
Comparing both studies will allow us to 'tune' up our system.

PAVCAS simulator

The problems we have to cope with are to put the driver into a sufficiently realistic situation in order to get a meaningful assessment, without however exposing him to any danger.
Taking into account the difficulties to carry out such a study in a real environment, it has been decided to implement a driving simulator: the PAVCAS simulator was born!

The design of this simulator directly depends on the experiment fields
The aim of this paper is to show how the PAVCAS simulator was designed in order to study human vigilance in a highway driving condition.

The challenge is to develop a simulator as realistic as possible to avoid measures distortions and flexible enough to modify easily simulation parameters.

In this paper, we try to show how we did account for these constraints to design the various parts of the simulators:
- simulator architecture
- the visual system
- the models
- the cabin with the motion rendering and the wheel steering actuator
- the sound system

1.SIMULATOR ARCHITECTURE

1.1 performance

We have to take care of the visual time response and of the motion time response of the simulator.

The visual time response is the delay between driver action and the simulator visual reaction.
It includes:
- sensors acquisition
- vehicle model computation
- image computation and display

The motion time response is the delay observed between the driver action and a motion reaction.
It includes:
- sensors acquisition

- vehicle model computation
- motion computation and control

The visual time response and the motion time response must be kept equal and below 100ms not to disturb the driver for physiological reasons (Ask Mr. Muzet at LPPE Laboratory for more details).

To achieve these performances we have implemented a parallel network distributed architecture where each task has a specific computer:
- a pentium PC for model computation and axis motion control, named 'PC Asservissement'.
- a Silicon Graphics reality engine2 ONYX for visual computation and display, named 'G.I.S.'.
- a PC to collect data and analyze driver behavior named, 'PC Analyse'.

These computers are interacting one another in two different ways:
• at first, at the beginning of the simulation, the data-base describing the virtual environment is copied to each computers.
• then, during the simulation, computers exchange information through buffers:
- 'command buffer' for the link between 'PC Asservissement' and 'G.I.S.'. It contains mainly the position of the vehicle.
- 'Analyse buffer' for the link between 'G.I.S.' and 'PC Analyse'. It contains traffic and road environment informations.

Figure 1 gives a general overview of the architecture described here.

1.2 modularity

The different parts can be disconnected and tested separately.

For each subsystem parameter in term of time-response, gain can be modified, allowing us to tune the system.

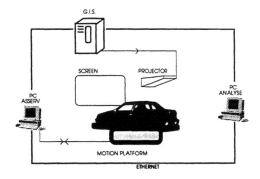

figure 1: simulator network architecture.

It is an open architecture:
- Other computers could be added to the network for a specific task or to upgrade the system.
- These computers could be also PC, work-station, specific physiologic recorders etc.

2.THE VISUAL SYSTEM: G.I.S. AND PROJECTION SYSTEM

2.1 performance

We call G.I.S.(Générateur d'Images de Synthèse) the graphical station

To avoid symptoms such as eye strain, headache,... we need a fluent image.
It means a 30 Hertz minimum frequency and anti-aliased images.
To avoid a flickering phenomenon, the screen frequency must be above 50 Hertz.
In addition, as we use steering wheel movements and lateral position as behavioral indicators we must be particularly careful with the time-response, and the realism of the road image. In this respect, we must use of textured and aliased pictures which improve the driver location on the road.

All these specifications imply to select a really powerful station:

figure 2: example of road environment.

The vision simulation is performed by a Silicon Graphics reality engine2 ARM ONYX 100Mhz (stabilized textured 30 i/s)

Projection is done by a multiscan projector SONY VPH-1272
The front view is 40° horizontal and 30° vertical.

Figure 2 displays an image of the road environment

2.2 modularity

The graphical station can be upgraded in terms of pixel-power and image resolution plugging 'Raster Manager' Silicon Graphics cards.

We have a road environment editor that enables us to generate easily, off-line, road data-base including panels,...
You can see an exemple of highway security panel on figure 3.

The road is build with predetermined road sections. You can extend the road environment with new types of road sections. For example, figure 4 illustrate the building of a ring road.

During the real time simulation, we can change some environment parameters:
- light
- fog

figure 3: example of panel.

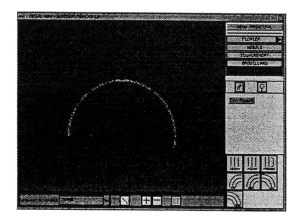

figure 4: building the road with the road editor.

3.THE MODELS

3.1 performance

One of the fundamental key of accuracy in simulating a real vehicle's behavior is the tire-road contact model. For example, if we do not model the slide of the tire, the driver reach lateral acceleration up to 2g while a common car lateral acceleration is below 1g.

Such a model involves:
- a real world roadway geometry model
- a model of real surfaces
- a model of deflecting and enveloping tire responses

No doubt, these three models are inter-related.

We must keep in mind that we are talking about dynamic real-time models (120Hz rate!)

3.2 modularity

It is easy to change the parameters for each models.
For example:
- road curves
- elevation difference across a tire's surface path.
- stiffness of contact tire-road

Each model is itself divided in several subsystems.

4. THE CABIN WITH THE MOTION RENDERING AND THE WHEEL STEERING ACTUATOR

4.1 performance

The real-time simulation of vehicle dynamics yields a significant amount of valuable data that a motion platform can directly render. It gives the driver a better knowledge on the behavior of his vehicle

However, we must take care of the consistency between movement and visual cues for the driver because it could be a source of simulation illness.

The natural approach should be to define a motion platform to reproduce vehicle accelerations. Such a platform would require very large displacement amplitude and power. This means a very expensive system. Considering that the driver is more sensitive to a variation of acceleration than to constant acceleration one, a cost

effective way is to render only acceleration derivative.

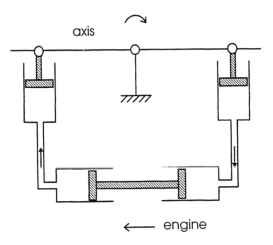

figure 5: principle of power hydraulic transmission.

This approach leads to a small but accurate displacement and a fixed image display.
The PAVCAS simulator uses the second approach.

The PAVCAS simulator has six motion axis:
- roll and pitch of the cabin to render lateral and longitudinal acceleration derivative
- roll and pitch of the driver seat to reinforce this acceleration and to tighten the safety belt giving a realistic sensation when slowing down quickly.
- a longitudinal axis of the cabin to reinforce longitudinal vehicle accelerations.
- a vertical axis of the cabin to render a 'pump effect'

Motion is issued by six brushless electrical engines. Motion is transmitted to axes through an hydraulic transmission (see figure 5) in order to avoid noise and motor disturbance. The use of such a transmission in this context is unique.
Engine power outputs are defined to provide accelerations corresponding to 5Hz with 10mm amplitude.
Engines are driven by power amplifiers providing direct speed control and engine position encoder outputs. These amplifiers are driven by an external motion control card. For better control, this card takes into

account the real position of each axes, thus compensating the distortion introduced by the hydraulic transmission.

4.2 modularity

Each axis can be disconnected separately. It enables us to tune each axis separately and to study the influence of each axis on driver performances.

In addition, we can include perturbation on each axis to study the driver reaction to it.

5.THE SOUND SYSTEM

The basis of sound simulation is a playback of recorded noises samples. The engine noise is mixed with a white noise which increases with speed thus simulating the aerodynamic noise.

The sound is performed by a customary sound-synthesizer. 'PC Asservissement' computer controls the frequency and amplitude of these sounds according to the vehicle model.

The treble part of the sound is transmitted to the car loudspeakers mounted on the front part while the bass part is transmitted to a bass loudspeaker mounted at the rear of the car.

CONCLUSION

The integrated simulator is now under an experimental phase in which a series of questions have been considered in terms of realism Vs our needs.

The result is a simulated environment in which the driver receives many cues providing us a large field of experiments. The PAVCAS research program is intented to show that a simulator could be a tool to study human behavior.

It should be a shame that the constant progress in the electronic, telecom, and computer field could not secure driving task.

ACKNOWLEDGMENTS

We are glad to acknowledge the precious contribution of the staff members of L.A.A.S. and L.P.P.E..
Of course, we gratefully acknowledge the partners who founded the project, I mean: la fondation MAIF, la Caisse Nationale d'Assurance Maladie, l'Association des Sociétés Françaises d'Autoroute, le Conseil Régional de Midi-Pyrénées and le Centre National de la Recherche Scientifique.

REFERENCES

DELTEIL J (april 1995). Linear networks modeling for driving simulation and ground vehicle. In: *ITEC 95 proceedings pp.*

DELTEIL J.(september 1995). Modélisation de réseaux linéaires pour l'exploitation du simulateur PAVCAS. In: *DSC'95 conference proceedings* (Neufs Associés), p 303-314, TEKNEA, Toulouse.

DESIGN AND IMPLEMENTATION OF EMBEDDED EXECUTIVE
FOR AN OBSTACLES DETECTION SYSTEM

N.E. Zergainoh, S. Bouaziz, T. Maurin

AXIS Lab., Institut d'Electronique Fondamentale, IEF
Université Paris-Sud, Bat. 220, 91405 ORSAY Cedex FRANCE
Phone (33-1)69-41-78-04, Fax (33-1)60-19-25-93, zergaino@ief-paris-sud.fr.

Abstract: In this paper, real-time distributed executive is presented which is suited for
embedded obstacle detection system. The real-time distribution and parallel execution
are supported by generic executive kernel which contains a portable set of services and
real-time micro-kernel which controls concurrent processes and schedules their
execution. The proposal executive is obtained by graph transformation methodology.
This approach exploits the inherent parallelism of the application, preserves the
formal specification of program and provides developers with control over the
evolution of an application during its development. Consequently, the executive
obtained is a compromise solution between portability and efficiency.

Keyswords: obstacle detection system, codesign methodology, real-time, distributed
executive, microkernel, portability, efficiency.

1. INTRODUCTION

All on board processing for real-time obstacles
detection and vehicle control in French Pro-chip
demonstrator[1] are handled by experimental
multiprocessor system with distributed memory
combining multiple high-speed communication links
and the specific communication networks used to
link the sensors to different nodes (Bouaziz, et al.,
1993). All sensorial information are transmitted over
the Vehicle Area Network, VAN. Two VAN types
are used, the fast VAN for transmitting sensorial
information provided by infrared and radar sensors
and the slow VAN permitting to connect others
equipment such as beacons, inner indicators or man
machine interface. The processing structure can be
roughly defined by three mains modules. Each
module is itself composed of several complex
algorithms, see (Reynaud and Maurin, 1994). The
key to success in such system is the timely execution
of data-processing tasks that usually reside on
different nodes and communicate with one another
to accomplish a common goal. End-to-end deadline
guarantees are not possible without a communication

network that supports the timely delivery of inter-
task messages, see (Zergainoh, et al., 1994b).
The design of such system is a difficult task. Several
factors tied to hardware and software jointly
determine system performance. Numerous
engineering tradeoffs exist between these
parameters. Historically, the design and development
of multiprocessor systems have been done at a low
level. The user is often faced with debugging the
algorithm, hardware and software implementations
simultaneously. The complexity and the safety
constraints of our application require safe
specification and implementation method, see
(Zergainoh, et al. 1994a) A method intended to
support the development cycle from conception
through specification, analysis and prototyping to
final implementation.

We propose a methodology, for mapping algorithm
into architecture, based on graph models as much for
specifying the algorithm (to exhibit the potential
parallelism) and the architecture (to exhibit the
available parallelism) that to describe the
implementation in term of graph transformations
(reduction of parallelism). The result of these
transformations is a distributed executive program
supporting the real-time execution of algorithm on
the target architecture. The mapping technique

[1]This reseach has been supported partly by the G.I.E. P.S.A.
R.N.U.R. under contract CNRS/GIE-PSA with the CNRS and
partly by a DRET/CNRS Contract.

consists in choose the implantation that respects at best the totality constraints of the real-time application (Sorel, 1994). SynDEx is an interactive software environment which implement this methodology (Lavarenne, *et al.*, 1991)

The remainder of this paper is organized as follows. Section 2 describes the methodology. Section 3 gives the architecture of our executive model. Section 4 presents the French Pro-chip obstacle detection system. In section 5, we describe the development cycle and the real-time implementation of the system using SynDEx environment. Section 6 summarizes the work.

2. METHODOLOGY FOR MAPPING ALGORITHM INTO ARCHITECTURE

The methodology is based on two graph models. Algorithms may be modeled by a data-flow graph model called a *software graph*. A multiprocessor is modeled by a hypergraph called a *hardware graph*. The implementation process consists of reducing the potential parallelism of the *software graph* into the available parallelism of the *hardware graph*.

The algorithm is described by a data-flow graph model where the vertices are actions and edges are data-flows. This description exhibits the potential parallelism of the algorithm. In practice, the algorithm is described using formal programming style in which a high-level synchronous data flow language SIGNAL (Benveniste and Le Guernic 1990). The multiprocessor is modeled by an hypergraph where the vertices are processors and the edges are interprocessor communication physical medium (point-to-point or multi-point). Each processor contains a memory, a computation unit, at least one interprocessor communication unit and some external I/O units.

The implementation is obtained by partitioning the software graph and by scheduling on each processor the computational actions which have been assigned to it. The partitioning and scheduling preserve the consistency of the synchronous specification. The inter-processor communication is assumed by a communication action which is inserted in the software graph between the computations. Each communication action determines the type and the destination of the transmitted data. Automatic mechanisms of scheduling cannot easily exploit particular properties of an application. Reliance on low-level facilities also compromises portability. Hence, we prefer an alternative approach, the partitioning and the scheduling optimization are expressed by simple control statements, which may be done either automatically or manually.

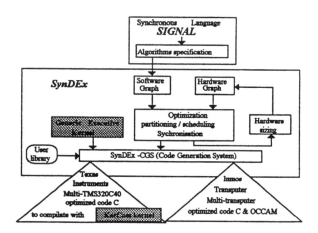

Fig. 1. Block diagram of SIGNAL-SynDEx software environment

The optimization problem consists in finding the mapping and the scheduling which minimize the response time. In our model, all task characteristics are known a priori. then the static scheduling is the most optimal scheduling. Many practical instances of scheduling algorithms have been found to be NP-complete, i.e., it is believed that there is no optimal polynomial-time algorithm for them. Since a feasible schedule is time consuming to find and we need to find a feasible schedule quickly, we take a heuristic approach. The heuristic scheduling algorithm tries to determine a full feasible schedule for a set of tasks in the following way. It starts at the root of the search tree which is an empty schedule and tries to extend the schedule, with one more task, by moving to one of the vertices at the next level in the search tree until a full feasible schedule is derived. To this end, we use a heuristic function H which synthesizes various characteristics of tasks affecting real-time scheduling decisions to actively direct the scheduling to a plausible path. The heuristic function is applied to each of the tasks that remain to be scheduled at each level of search. The task with the smallest value of function H is selected to extend the current schedule.

The real-time distributed executive, implemented as a source-to-source transformation, takes application program, control statements, scheduling routines and communication actions of the generic kernel and generates a new parallel program that must be compilated with the fast micro-kernel.

3. DISTRIBUTED EXECUTIVE MODEL

The proposal model divides the software into three levels: arithmetic processes, communication processes and interface level.

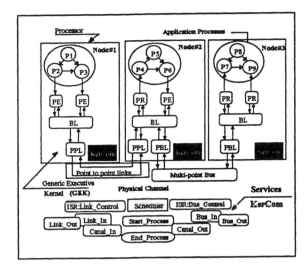

Fig. 2. Conceptual model with possible mapping into a Topology.

The application processes are executed in the highest software layer. An application is composed of a set of processes and cooperates with the other application processes via the objects of the generic kernel executive, GEK.

The objects of the generic executive kernel GEK are found in the intermediate layer. These objects offer interprocess communication to the application processes and they are constructed using elementary primitives offered by the microkernel KerCom. The objects of GEK are independent of the target.

The lowest level is the interface to the hardware level and is composed of the microkernel, KerCom. KerCom is composed of a number of elementary objects permitting the construction of objects of much more complex models. KerCom is the heart of the system and permits implementing the model on the target machine.

This model has several others advantages compared to the others approaches. It permits programs generation that are portable over a wide range of multiprocessor machines. This approach separates the application processes from the system process which are themselves separated into a portable high level and non portable low level. This separation provides maximum portability and preserves the efficiency of the executive. (see Fig. 2).

3.1 Generic Executive Kernel

The generic executive kernel GEK is the set of objects permitting the construction of real-time distributed executive. The objects of GEK insure a message synchronization, supports nearest-neighbor communications and handle message routing

between nodes which are not direct neighbors, these simplified the porting process and also minimized the amount of code present on each node (static structure). There are 4 types of objects. The sender object called PE (Porte Emission) is connected to an output activity of the algorithm graph whose role is to format messages and add routing information to the outgoing messages. The receiver object called PR (Porte Reception) is connected to a receiving activity. The role of the PR is to unformat messages, and to transmit them to the receiving activity. The software bus called BL (Bus Logiciel) is connected to all objects and its role is to route messages and to implement flow control. The Input/Output object called PES (Porte Entree/Sortie) insures the physical transfer of interprocessor messages. There are two types of PES: PPL (Processor to Processor Link) and PBL (Processor to Bus Link).

3.2 The Microkernel KerCom

KerCom is at the heart of the system. It is invoked each time there is a modification in the set of active processes. It assures the management of process switching. KerCom chooses a process from the set of active processes while respecting the priority ordering of processes and garantees a time latency (in order to meet the real-time constraints). It offers a unique elementary communication mechanism permitting the constructing of complexes communication using the generic kernel, GEK. The communications mechanism is based on message passing and utilizes the concept of rendez vous for process synchronization. This interaction is assured by the communications channels. This technique permits the synchronization of concurrent processes on the one hand and solves the problem of shared resources (that can possibly create deadlock) on the other. KerCom centralizes and resolves the problem of mutual exclusion by proposing two elementary primitives for communication between two processes on the same processor and two other primitives for communication between two processes executing on different processors using the same principle. The communication channels are realized by using memory allocation for the internal communication and using point to point links or multipoint links (physical bus) for external communication.

Real-Time Scheduling. In our model, each processor executes an arithmetic process (each of which is composed of several sequential arithmetic processes) and several concurrent communication processes (objects of GEK). The processor executes the arithmetic processes based on their corresponding process priorities. The process currently executing is called the current process and the set of processes that are ready for execution is called the set of active processes. If the current process is a high priority

process it continues its execution until it terminates or is blocked. There are three way a process may be blocked:
- By waiting for a input channel, the current process is waiting for data from the channel that the sending process has not yet transmitted.
- By waiting for an output channel, the current process is trying to transmit data on the channel but the receiving process is not yet ready to receive the data.
- A process becomes active whose priority is superior to the current process.

Kernel primitives. KerCom offers a set of elementary primitives permitting the constructing of GEK objects. The start and termination of concurrent processes is assured by the *Start_Process* and *End_Process* primitives. The interprocess communication in the same processor is assured by the *Canal_In* an *Canal_Out* primitives. The processors have a communication link interface assuring the physical transfer of data from one processor to another. They are connected by the point to links. KerCom offers two primtives for interprocessor communication *Link_In* and *Link_Out*. The multi-point communication (via the physical bus) is offered by *Bus_In* and *Bus_Out* primitives. The implementation of these primitives is tight linked on the one hand with the type of physical bus used and on the other hand with the communication protocol. we have implemented these primitives for special case of the physical bus such as the Vehicle Area Network (VAN bus). For establishing one communication on a set of available communication channels, KerCom offers the alternative primitive called *Alt_Process*. It permits implementation of the BL object of GEK.

4. OBSTACLES DETECTION SYSTEM

The goal of the application is to give assistance to vehicles drivers. The project aims to introduce custom hardware intelligent processing. The design of such system is a difficult problem which has to cope with numerous constraints such as response time and reliability. An effective situation analysis relies on different kinds of sensors providing rough data. These data are used to extract pertinent numerical information. these information are then analyzed to reconstruct environment, to forecast danger and to prevent collision. Fig. 3 illustrates the obstacles detection system. All on board processing for real-time obstacles detection and vehicle control are handled by experimental multiprocessor system with distributed memory combining multiple high-speed communication links and the specific communication networks used to link the sensors to different nodes.

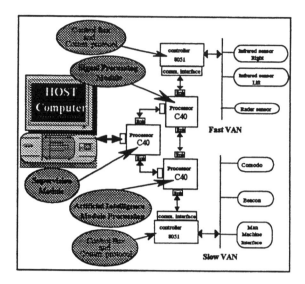

Fig. 3. Obstacle detection system

The processing is supported by 3 Texas Instruments TMS30C40 parallel processor. The TMS320C40 processor combines six communication ports with multi-channel direct memory access (DMA), running in parallel with the CPU.

All sensorial information are transmitted over the Vehicle Area Network, VAN. Two VAN types are used, the fast VAN for transmitting sensorial information provided by infrared and radar sensors and the slow VAN permitting to connect others equipment such as beacons, inner indicators or man machine interface. The controllers of the 8051 Intel type are used to handle the VAN communication protocol as well as the interface of VAN bus with sensors and computing nodes. The host computer of the IBM PC type serves for mass storage, software development and perception front-end machine.

The processing structure can be roughly defined by three mains modules. Each module is itself composed of several complex algorithms.

M1) Signal Processing Module: Provides the first reconstruction of the scene from two scanning IR sensors by use of triangularization. The sequence of processing is noise filtering, matching step between events coming from left and right sensors, reconstruction and estimation of the scene by Kalman filtering.

M2) Intelligence Artificial Module: Digital to symbolic information conversion and decision making on rules based expert system are used to analyze the driving situation.

M3) Supervision Module: Control and verify the coherence of the pertinent parameters for each process and provides diagnostic information.

5. IMPLEMENTATION AND RESULT

We have parallel system submissive to strict timing constraint and the severe reliability constraint (indeed the functioning failure of system could lead to catastrophe). For this reason, our system requires safe design methods. An executive is essential to guarantee all constraints required by the application. The use of formal specification language in the software development process is the object of our methodology. High-level specifications are gradually transformed into executive code in a programming language.

One of the central ideas behind our executive model is processor and topology independence. This was achieved through the use of the C language. The GEK has been written with C language and KerCom has been written for the major part with optimized C. Only the specific operations such as task switching and interrupt handling were written in assembly language. When the kernels are ported to new processor, only the processor specific operations have to be redesigned. Hence the kernels can be ported to most popular processors in few weeks. The single communication model supports systems with different types of interprocessor communication media such as link and bus. The hard real-time characteristics are guaranteed by using priority ordering at all times. The message based protocol is powerful partly by its portability but also because of its simplicity and elegance. It is very secure as it isolates all memory operations from one another.

Our application necessitates deep study of both architectural and algorithmic issues. The SIGNAL-SynDEx software environment provides tools to develop this kind of application. A SIGNAL program specifies an algorithm. The SIGNAL compiler mainly consists of a formal system which is able to process signals clock, logical time and dependency graph. SynDEx takes as an input the code description by the SIGNAL compiler of the algorithm specification provided correct, thanks to the synchronous compiler, performs performance evaluation of distributed implementations, propose and optimized distribution, and automatically generates code of multiprocessor implementation. We describe the hardware application as SynDEx hardware graph and the software application as a SynDEx software graph, then map the two graphs by specifying placement constraints. Then we run a code generator which produces the multiprocessor executive. The Code Generator System (SynDEx-GCS), implemented as a source-to-source transformation, takes application code, control statements, scheduling routines and objects of generic kernel and generates a new parallel program that must be compiled with KerCom kernel. Fig. 4 shows SynDEx hardware graph and the

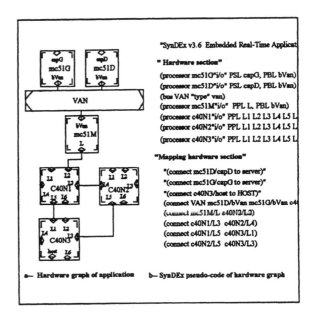

Fig. 4. Application description

corresponding pseudo-code of the hardware architecture of the application. Fig. 5 illustrates the VAN protocol model and how SynDEx inserts the distributed communication executive in this protocol, and Fig. 6 shows an application software graph mapped on the hardware graph.

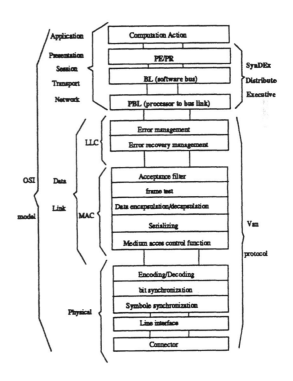

Fig. 5. Distributed executive in VAN

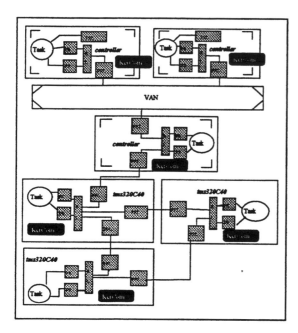

Fig. 6. Vehicle application implementation

6. CONCLUSION

The multiprocessor solution in our real-time application improves the overall performance of the system in two different ways. Obviously it improves performance when parallel processing tasks are performed but it also improves performance when repetitive sequential tasks are performed. This second improvement is achieved by a partitioning according to a pipeline organization. Parallel processing requires processor synchronization, parallel task partitioning and shared resource access.

The SIGNAL-SynDEx tools provide support for parallel processing implementation. We have presented an optimized real-time distributed executive for obstacles detection system. This executive is obtained by graph transformation methodology of formal specification. This method separates the specification of application from its implementation. It provides developers with control over the evolution of an application during its development. The development cycle time is tremendously reduced as well.

Currently, we have an evolutive real-time obstacle detection system. We have tested the totality functioning of the chain going from sensors until the alarm to the pilot. We have elaborate two methods (analytic and experimental) for a complete evaluation of performance. This evaluation provides complete time latency analysis of the different modules, and an evaluation of the executive cost impact.

ACKNOWLEDGMENTS

The authors wish to express their gratitude to Yves Sorel and Christophe Lavarenne in the project sosso INRIA for the ideas generated through theirs many interactions with them.

REFERENCES

Benveniste, A. and P. Le Guernic. (1990). Hybrid dynamical systems theory and the language SIGNAL. *IEEE Transactions on Automatic Control.* **Volume 17 No 6, pp. 535-546.**

Bouaziz, S. R. Reynaud and T. Maurin. (1993). Parallel architecture for an embeded real-time application. *ICSPAT, Santa clara USA.* **Volume 1, pp. 151-155.**

Lavarenne, C., O. Seghrouchni, Y. Sorel and M. Sorine (1991).The SynDEx software environment for real-time distributed systems design and implementation. *Proc. of the European Control Conference.* pp. 535-546.

Maurin, T. (1994) "The French PRO-CHIP Demonstrator for obstacle detection and avoidance", 3rd Prometheus Workshop on collision avoidance, June 1994, Stuttgart. Proc. PROMETHEUS.

Reynaud, R. and T. Maurin (1994). On Board Data Fusion and Decision System Used for Obstacle Detection: 2D vision Versus 1D Sensor Fusion. *IEEE International Symposium on Intelligent Vehicles.*Paris, **FRANCE**, pp. 545 550. January, 1994.

Sorel, Y. (1994). Massively Parallel Computing Systems with real-time constraints. *Proc Massively Parallel Computing Systems,.* **Italy**, May 1994.

Zergaïnoh, N.E., T. Maurin, Y. Sorel and C. Lavarenne (1994a). A real-time multiprocessor development environment: design and implementation. *Euro94 Workshop on Parallel and Distributed Processing, IEEE Computer Society.* **Malaga**, pp. 545 550. January, 1994.

Zergaïnoh, N.E., T. Maurin and R. Reynaud (1994b). "Synchronous Protocol for Real-time Communications in Intelligent Vehicle," *IEEE International Symposium on Intelligent Vehicles.* **Paris, FRANCE**, pp. 545 550. January, 1994.

Towards A Robust Algorithm for Dynamic Obstacles Tracking

Kamel BOUCHEFRA [1,2]   Roger REYNAUD [1]   Thierry MAURIN [1]

[1]Institut d'Electronique Fondamentale
Université Paris-Sud
91405 Orsay
France

[2]Ecole supérieure d'Ingénieurs en Informatique et Génie des Télécommunications
1, rue du port de valvins, 77215 Avon-Fontainebleau cedex

Email: kamel,reynaud,tm@ief-paris-sud.fr
tel : 69-41-65-74 / 69-41-78-04
fax : 60-19-25-93

Abstract

This paper deals with the management of road traffic collision risk at the software level of vehicle on-board systems. The proposed algorithm consists in two processes and aims to carry out obstacle tracking. The study is based on a multiagent approach. Each process of the algorithm is an agent. The first agent deals with sensor data and the second one deals with data either issued by the first agent or obtained by exchanged communications with the equivalent agent of another vehicle. The second agent achieves the tracking over the time of protagonist obstacles. The main point in this paper relates to the interactions among agents within a same vehicle. The dependencies that exist among the agents are associated to a level of satisfaction of the tracking agent.

Keywords: dynamic obstacles; sensors data handling; tracking obstacle automaton; agents; interactions.

1 Introduction

The management of road traffic is currently the subject of research programs in the main industrialized countries [14, 3, 10, 15, 2, 6, 8]. These studies aim to improve road traffic security and manage its actual state and its future evolutions. As part of the research done in the field road traffic systems, the arising aspect in our research is concerned with the management of the collision risk by an on-board collision avoidance system [1, 11]. The dynamics of road traffic is the source of the uncertainty which an on-board system must deal with and reduce. Observing the universe and obtaining information about its state may reduce the amount of uncertainty. In this study, the universe of discourse is restricted to the protagonist vehicles. So observing the state of the universe is equivalent to observing the protagonist vehicles and in this frame of mind requires a robust obstacle tracking algorithm. Such an algorithm is described in this paper.

This study supposes that some vehicles of the road traffic system have on-board systems with perception, communications and calculations capabilities (a vehicle with such capabilities is called *the observant vehicle* in the paragraphs below). These capabilities allow the classical perception, processing, action scheme.

The first paragraph of this paper presents the architecture of our simulator of an on-board system dedicated to road traffic collision avoidance. This study is based on a multiagent approach [7, 16, 9]. The agents are the dedicated processes of the on-board system. The emphasis is on the obstacle tracking algorithm consisting of two processes. One process is situated at the sensor level, it consists in data handling and is achieved by the sensor agent. The results of this step are an entry of an obstacle tracking process devoted to the so-called tracking agent. Another entry to this agent are data exchanged with a protagonist vehicle by use of communication capabilities. This last way of getting data involves two protagonist vehicles both having vehicle to vehicle communication capabilities. Also, this may be of great interest for the algorithm robustness [8, 11]. The second paragraph describes these two agents. The third paragraph presents the interactions between the agents. These interactions are associated to a level of satisfaction of the tracking agent. In our scheme, this agent aims to become satisfied by breaking the dependency relation that naturally associates it to the other one. The tracking agent is in a high level of satisfaction when it knows all of the protagonist obstacles. In such cases, more facility may be expected for higher level processing consisting in situation analysis (see §2) in order to improve safety and comfort of travel [11].

2 The architecture of the simulator

The design of real-time systems leads finally to hierarchical schemes for the organization of the processing [5, 13]. This allows systems which have both reactive and cognitive behaviours. To realise these schemes, one has to identify the different levels of processing that are involved. This was the basic idea in our design of the simulator of an on-board system.

In our study, there are three levels of processing: 1) the processing at the low level consisting in obtaining and processing data representing the universe state. The tracking algorithm presented in this paper is situated at this level of processing. 2) the processing at a mid level consisting in local analysis of the universe state. Conflicts among the vehicles of the road traffic are identified at this level. 3) the processing at a high level consisting in semantic and global analysis of the universe state. Conflicts among vehicles are solved at this level.

The processes at the different levels are involved to carry out different behaviour schemes. These are the following: 1) the reactive scheme, it associates processes of the first two levels. The reactive behaviour is expected when the system has to deal with hard time constraints. 2) the planification scheme, it also involves processes of the first two levels and is expected when the constraints within the system are solved and when time constraints are not hard. 3) the decision scheme involves the processes of the three levels when the system deals with a universe state characterized by conflicts that are not solved and when time constraints are not hard.

The emphasis is on systems that produce answers to universe stimuli with the state of the universe being estimated as well as possible and for as long as possible according to time constraints.

Our first step towards the design of such a system is the algorithm described in the remaining paragraphs.

3 The sensor agent

In [4] there is a detailed description of a processing scheme at the sensor level. To carry out such an algorithm is not our interest in this study. Indeed, we deal with sensors from the behavioural point of view.

Different kinds of sensors may be used, each kind having its specific characteristics. For the observant vehicle, a whole system based on 4 sensors is supposed. There are two sensors among the four that are supposed to allow the localisation of each perceived obstacle. Processing at these sensors level consists in what is known as a triangulation process [1]. The information in this case is considered uncertain and accurate in a satisfactory manner (see 3.1). One may deal with such information when using infra-red based technology sensors [12]. These sensors are called S1 and S2 sensors hereafter. The remaining sensors are supposed to allow the localisation and estimation of the velocity of each perceived obstacle. The information in this case is considered certain and inaccurate (see 3.2), one may deal with such information when using for example laser telemeter sensors. Each one of these sensors processes independantly of the other. These sensors are called S0 and S3 sensors in the next paragraphs. All of the sensors observe the same scenery([1]).

The hypothesis of using two kinds of sensors allows to pass by the uncertainty and the inaccurate nature of the information delivered by each of the sensors models. It also allows some robustness of the agent (see §3.3.).

In order to achieve the data processing of this level, the process to take place, must take into account the use of various sources of data; so the proposed algorithm achieves data handling in a process with cascadability and parallel processing capabilities as shown in figure 1. Each node within the graph represents a task achieving one specific process. These tasks are noted T1, T2, T3 and T4 and are described hereafter.

3.1 Description of task T0

Task T0 carries out a triangulation based process of data issued from S1 and S2 sensors. This results in a set of possibilities for obstacle locations. However, each possible location that is detected is known with a satisfactory accuracy. The cardinal of the set of all possible obstacle locations is equal to $\sum_{i=0}^{n-1}(n-i)$ where n is the number of obstacles. In figure 2 is shown the set of ten

---

[1] Let us note Ω the observed scene.

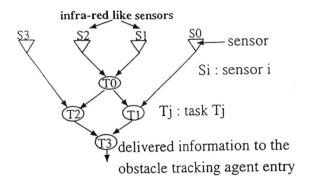

infra-red like sensors

S3 S2 S1 S0 — sensor

Si : sensor i

T0

T2 T1 Tj : task Tj

T3 delivered information to the obstacle tracking agent entry

Figure 1: data flow graph of the sensor agent's process

possibilities of obstacle locations in the case of 4 detected vehicles.

The relation that associates n perceived obstacles to $\sum_{i=0}^{n-1}(n-i)$ possible locations represents the uncertainty that has been mentionned previously.

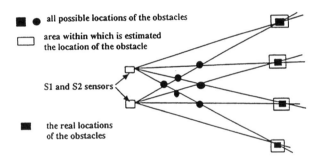

Figure 2: uncertainty of information issued by S1 and S2 sensors

Let V_j, $V_j \in \Omega$ be the jth possible location of the perceived obstacles; then Task T1 results in the set $PossibleLocations = \{ V_j \setminus j \in [0, M]$ where $M = \sum_{i=0}^{n-1}(n-i)$, n is the number of perceived obstacles and V_j is the jth possible location associated to obstacles$\}$.

3.2 Description of tasks T1 and T2

Tasks T1 associate the results of Task T0 with data from sensors S0. Task T2 processes the same as Task T1 with handling data from sensor S3.

The set $CertainAreas$ is defined as the set of all areas[2](see figure 3) considered in a manner that

---
[2] Where area is a subset of Ω within which is estimated the location of an obstacle.

is certain by S0 and S3 sensors as containing an obstacle:

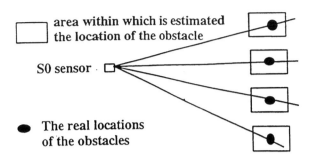

Figure 3: inaccurate nature of data issued by S0 and S3 sensors

Task T1 results in the set $CertainLocations = \{ V_j \setminus j \in [0, n]$ where n is the number of perceived obstacles and $V_j \in PossibleLocations$ and $V_j \in CertainAreas\}$.

The algorithm of task T1 is given below:

```
n = number of perceived obstacles
for i = 0 to ∑_{i=0}^{n-1}(n − i),
    if V_i ∈ (PossibleLocations ∩ CertainAreas)
        then V_i ∈ CertainLocations
    if (Card(CertainLocations) == n)
        then exit
end for
```

3.3 Description of task T3

Task T3 carries out a supervisory process by comparing data from tasks T1 and T2. The results obtained by one task (say T2 for example), must be the same as those obtained by the other (T1). Task T3 also allows robustness of the perception process as redundant sensors are used.

A satisfactory processing of this task is obtained when only sensors (S0, S1, S2) or sensors (S3, S1, S2) are functionning. A less happens in the cases where only S0 or S1 or (S1 and S2) sensors are functionning.

A critical situation for the agent happens only when (S0, S3 and S1) or (S0, S3 and S2) sensors are broken simultaneously.

Task T3 transmits its results to the tracking agent.

4 The tracking agent

The process devoted to the tracking agent consists in associating each perceived obstacle to its location and velocity in a process taking place over time. This operation is translated in our case through descriptive cards, each one associated to a specific obstacle. Each card exists for the time the obstacle remains perceivable.

This agent knows the existence of an obstacle through the processing steps made by the sensor agent or by its own exchange of communications with the equivalent agent of a protagonist vehicle.

The agent has one entry point at the sensor agent. Thanks to this entry point the agent can act on the sensor agent.

Our design of this agent consists in a four state automaton - see figure 4. Each state (state0, state1, state2 and state3), measures the quality of the available information. State 0 deals with each obstacle whose location and velocity remain undetermined. State 1 deals with each obstacle whose location and velocity are determined but for which the agent wishes some confirmations in the ulterior perceptions. State 2 deals with each obstacle whose location and velocity are determined in a manner that is certain. State 3 represents each obstacle that has disappeared from the observant vehicle's field of perception; the associated card is destroyed after some latency time that represents our way of verifying that the obstacle has really disappeared.

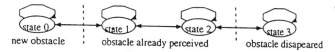

new obstacle | obstacle already perceived | obstacle disapeared

Figure 4: The tracking agent

5 Interactions among agents

Different kinds of interactions exist among our agents. These interactions reflect the different levels of dependencies of the tracking agent relative to the sensor agent.

These dependencies are represented as levels of satisfaction of the tracking agent. The agent aims to be satisfied : that means he wishes to know all of the detected obstacles. A high level of satisfaction of the agent is related to cases where all the obstacles are estimated in state2 or state3 of the automaton.

Interactions among our agents are:

- No interference and just observing interaction: this happens when the tracking agent depends entirely on the sensor agent in order to detemine the location and velocity of an obstacle. Such a situation represents a non satisfying level for the tracking agent. It corresponds to an obstacle estimated in state0 and to a level LS0 (see figure 5), of satisfaction for the agent.

- interference interaction: this happens when the tracking agent wishes[3] to correct the sensor agent's calculations in order to give more accuracy to the information. In such situations, the tracking agent already owns information that is certain thanks to previous processes or to exchanged communications with the tracking agent of another vehicle. This means that the protagonist vehicle has communications capabilities. Informations get by this way are known to be more accurate[11].

- No interference and not observing interaction: this happens when the tracking agent doesn't depend on the sensor agent in order to determine the location and velocity of an obstacle. The information is either known in a manner that is certain thanks to previous processes or is known thanks to exchanged communications with the tracking agent of another vehicle. If all the obstacles are estimated in state2 or state3 then the agent passes to its maximum level of satisfaction LS2.

These interactions are summerised in figure 5.

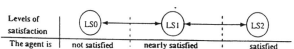

Levels of satisfaction | LS0 | LS1 | LS2
The agent is | not satisfied | nearly satisfied | satisfied

Figure 5: Different levels of satisfaction of the tracking agent

Levels of satisfaction are associated to the estimated states of obstacles.

Level LS0 is related to state 0, level LS1 is related to state 1 and level of satisfaction LS2 relates to state 2 or state 3. Transitions from one level of satisfaction to another are associated to transitions occurring at the automaton level.

Interference of the tracking agent with the sensor agent's calculations are allowed with regard to time requirements. If the dynamics of the

---

[3]The agent aims to improve its state of satisfaction.

road traffic allows it, modifications may take place. Otherwise (if a critical situation occurs, that means when the possiblity for an obstacle to become dangerous is too high), such modifications are forbidden.

The interactions among the agents are translated into rules among which are the following examples:

rule1 :
if   *the sensor agent estimates obstacle o_i in state0*
if   *the tracking agent hasn't initiated a communication with the perceived vehicles in order to know more about obstacle o_i.*

then   *the tracking agent is in an LS0 level of satisfaction, the tracking agent deals with the actual information known about obstacle o_i, the tracking agent initiates communications with the perceived obstacles in order to know more about obstacle o_i*

rule2 :
if   *the sensor agent estimates obstacle o_i in state0*
if   *the tracking agent has initiated a communication with the perceived vehicles in order to know more about obstacle o_i*

then   *the tracking agent is in an LS0 level of satisfaction, the tracking agent deals with the actual informations known about obstacle o_i, the tracking agent looks for an answer from the obstacle o_i*

rule3 :
if   *the sensor agent estimates obstacle o_i in state0*
if   *the tracking agent receives information about obstacle o_i*

then   *the tracking agent deals with the actual information known about the obstacle o_i*

6 Simulation

The processes presented in the previous paragraphs are the matter of a language oriented object simulator. This simulator allows instanciation of vehicles with the perception, communications and processing simulated capabilities. The processes communicate between them thanks to the cards introduced below. Vehicle to vehicle communications stand for 2 simulation steps while results of the other processes are obtained after one simulation step. Finally, a vehicle is described as follow:

```
class Vehicle {
communicationAgent();
sensorAgent();
trackingAgent();
decisionAgent();
planningAgent();
resultSaving();
}
```

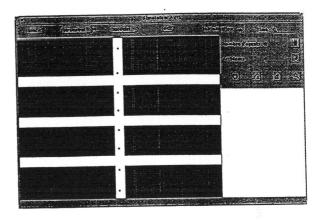

Figure 6: AVIS simulator

Each vehicle is given a path in such a way as to make vehicles almost meet in the same area at almost the same time. The paths of the vehicles are therefore conceived in order to maximize the number of interactions between vehicles.

The aim of the simulator is to validate the agents interaction scheme. Such interaction capabilities are of great interest. Indeed, the ability to use various sources of information allow the tracking agent to correct if necessary information obtained thanks to either a sensor agent's calculations or the execution of a communication step. Also, once an obstacles is well characterized (its location and speed are well defined), the sensor data handling stage can be overtaken for specific obstacle if a critical situation occurs.

The following table summarize the simulations that have been processed with the results that have been obtained after a same number (five in this case) of iterations for each situation of the sensors and communication processes. The first four columns indicate the state of each sensor during the simulation held. S0 being in state 0 means that the sensor S0 is not functioning. The fifth column indicates the fact that data have been obtained thanks to a communication step. The other columns indicate simulation results about respectively the state of the automaton and the tracking obstacle's level of satisfaction. Each line indicates the parameters of the simulations.

105

Simulations results						
SO	S1	S2	S3	comm	auto	satis
0	0	0	0	1	state 1	LS1
0	0	0	1	0	state 0	LS0
0	0	0 ·	1	1	state 1	LS1
0	1	0	0	0	state 0	LS0
1	1	0	0	0	state 0	LS0
1	1	0	0	1	state 1	LS1
0	1	1	0	0	state 1	LS1
1	1	1	0	0	state 2	LS2
1	1	1	0	1	state 3	LS2

table 1 : simulations results

7 Conclusion

The importance of using multiple sources of information is a well established claim. That amount of research in the field of data integration may be sufficient to prove it. Also, for a real-time system to behave with some robustness, it must successfully deal with the uncertainty that is specific to its environment.

In this paper, we have proposed an algorithm that is specific in that the utilisation of and the search for information about the environment is related in some way to the amount of uncertainty that is associated to the universe state: if the tracking agent is satisfied, it may ignore data issued from the sensor agent.

According to this work, our current research interest lies in a search for a loop involving a process and its surrounding environment in such a way as to produce guided interactions.

References

[1] A. Chebira R. Reynaud T. Maurin D. Berschandy. On board data fusion and decision system used for obstacle detection: A network and a real time approach. *Workshop On Real-Time Systems - Euromicro'91 Proceedings*, 1991.

[2] D. Bullock and C. Hendrickson. Roadway traffic control software. *IEEE Transactions on Control Systems technology*, 2(3), September 1994.

[3] C.W. Chen. Navigation of an autonomous land vehicle. *Proceedings of the 1987 IEEE International Symposium On Intelligent Control*, 1987.

[4] A. Chebira R. Reynaud G. Demoment. Fusion de données multicapteurs: application à la détection d'obstacles en temps réel. 13^{ieme} *Colloque GRETSI - Juan-Les-Pins*, September 1991.

[5] B. Chaib draa et E. Paquet. Routines, situations familières et non-familièrs dans les environnements multiagents. 9^{ieme} *Congrès RFIA*, January 1994.

[6] R.E. Fenton. Ivhs/ahs: Driving into the future. *IEEE Control Systems*, 14(6), December 1994.

[7] I. A. Ferguson. Touringmachines: Autonomous agents with attitudes. *IEEE Computer*, May 1992.

[8] T. Joubert and A. Kemeny. Cooperative driving: A distributed application. *The Real-Time Systems Conference, RTS'94*, January 1994.

[9] L. Kuo-Chu and al. A framework for controlling cooperative agents. *IEEE Computer*, jul 1993.

[10] M.A.Turk and al. Vits-a vision system for autonomous land vehicle navigation. *IEEE Transaction on Pattern Analysis and Machine Intelligence*, 10(3), May 1988.

[11] U. PALMQUIST. Intelligent cruise control and roadside information. *IEEE Micro*, February 1993.

[12] A.Chebira R.Reynaud, T.Maurin. Capteurs infra-rouge pour la reconstruction 2d modélisation markovienne et filtrage binaire. 14^{ieme} *Colloque GRETSI - Juan-Les-Pins*, September 1993.

[13] M. Morin S. Nadjm-Tehrani P. Osterling E. Sandewall. Real-time hierarchical control. *IEEE Software*, September 1992.

[14] S. E. SHLADOVER. Research and developpement needs for advanced vehicle control systems. *IEEE Micro*, February 1993.

[15] J.K. Hedrick M. Tomizuka and P. Varaiya. Control issues in automated highway systems. *IEEE Control Systems*, 14(6), December 1994.

[16] F. von Martial. *Coodinating Plans of Autonomous Agents*. Springer-Verlag, Berlin Heidelberg, 1992.

STATISTICS METHODOLOGY FOR AUTOMATIC DETECTION VIGILANCE IN REAL TIME: AUTOMOTIVE APPLICATION

**J. Armando Herrera-Corral, Said Labreche,
Neil Hernández-Gress, Daniel Estève**

*Laboratoire d'Analyse et d'Architecture des Systèmes
du Centre National de la Recherche Scientifique (LAAS - CNRS)
7, Avenue du Colonel Roche 31077 Toulouse Cedex, France
tel : (+33) 61.33.62.46 et 61.33.64.12; fax : (+33) 61.33.62.08
E-mail : herrera@laas.fr, labreche@laas.fr, hdez@laas.fr*

Abstract: Once the in board smart multisensor danger detection copilot system defined (1992). LAAS-CNRS has investigated different methodological approaches to realise the diagnosis. Three complementary directions have been identified : *a*. The following of the driving mistakes in terms of traffic regulations (1992); *b*. The evaluation time of the reactivity capability of the driver. We have studied Autoregressive Moving Average (ARMA), (1993) and *c*. The driving procedural analysis on which we are working in this contribution (1994). We are now planning to introduce a pre-metric by Principal Component Analysis (PCA). *Experiment* : A driver realises several times the experiment which consists of negotiating a cross-roads.

Keywords: PCA, Clustering, Metric, Correlation, Exploration

1. INTRODUCTION

Once the in board smart multisensor danger detection copilot system defined, (Martinez, 1992). LAAS-CNRS has investigated different methodological approaches to realise the diagnosis. Three directions to evaluate the driving quality have been identified:

a. Traffic Rules : The following of the driving mistakes in terms of traffic regulations. The diagnosis can be easily solved by expert systems or neural network systems (Chan *et al.*, 1992; Herrera, 1992; Herrera *et al.*, 1992).

b. Driver Reactivity : The evaluation time of the reactivity capability of the driver. We have studied Autoregressive Moving Average (ARMA) method on RNUR data related to the steering wheel movements (Chan and Herrera, 1993). We have detected some deviations between different drivers, but due to the non-stationary phenomenon, it isn't possible to have final and robust diagnosis.

c. Driving Procedures : The operational modes analysis on which we are working in this contribution (Herrera *et al.*, 1994). The problem is to identify the driving procedures (habitude) and its change in precise situations: cross-roads, overtaking...The hypothesis we have is that the changing status of the driver is accessible when the procedures vary. The behavioural change diagnosis can be realised by various methods, and principally by neural network. But some difficulties have appeared: Robustness, precision so, we are now planning to introduce a pre-metric by Principal Component Analysis (PCA).

Experiment : A driver realises several times the experiment which consists of negotiating a cross-roads. Several parameters are observed : - the speed of the vehicle, - the distance of the vehicle to the cross-roads, - the accelerator pedal, - the brake pedal, - the clutch pedal. - the indicator, - the gearbox, the steering wheel, etc.

Data : A series of data are then obtained with several drivers negotiating the m meters before the cross-roads.

Method : We process these data divided in intervals, to reduce working space in two or three dimensions, in order to save the maximum information by applying principal component analysis and classification method. So that similar data are classified interval by interval in a class. Knowing their origin, this classified data base can be used as learning data base to be exploited in Neural Network for generalisation use where unknown data are processed and results given according the memorised classes.

Principal Component Analysis (PCA) : Principal Component Analysis substitutes few synthetic variables called Principal Component for initial variables. A sub-group of initial variables is associated with each Principal Component by way of correlation (Benzecri, 1973; Cailliez, 1976). This analysis characterises and classifies the variables in a multisensor framework. The application considered an analysis of a driver's behaviours related to: the speed of his vehicle (v), his distance to the obstacle (d), the accelerator pedal (a), the brake pedal (b), the clutch pedal (c). The analysis was related to a distance to the obstacle.

Classification : However the PCA is one descriptive method and it doesn't always give significant partitions for the driver's behaviour. For this reason, we have used the classification method which maximise the between-group dispersion. Some parameters of this method (The metric, the number of classes) can be adjusted by the result of PCA (Celeux *et al.* 1989). The classification method will define the different driver's behaviours and the evolution of each behaviour related to the obstacle.

2. PRINCIPAL COMPONENT ANALYSIS (PCA)

Given a set of n elements $I = \{i/i = 1,...,n\}$ described by a set of p variables $J = \{x^j/j = 1,...,p\}$. We associate to this set a set of points $N_x = \{x_i/i = 1,...,n\}$ in a p dimension space. Both the two sets define data table X, where $X(i,j) = x^j(i)$ is the value taken by the variable x^j on the *i-th* element.

2.1 General Remarks

a. Metric in Subjects Space

The set of points is in a space $E = \Re^p$ called subjects space. We define on E a metric M associated to a scalar product M. The M metric allows to measure the 'similarity' between two points x and y belonging to E.

$$M(x, y) = \|x - y\|_M^2 = {}^t(x - y)M(x - y)$$

where with a super-script t, denotes transpose.

b. Metric in Variables Space

The set of variables is in a space $F = \Re^n$, called variable space. A weighting factor $w_i \rangle 0$ is associated to each element i so that $\sum_{i=1}^{n} w_i = 1$. It results a metric called weighting metric on F and noted D. The matrix D is a diagonal matrix so that $D(i,i) = w_i$. In the case of centred variables $\{x^j\}$.

$$\text{var}(x^j) = D(x^j, x^j) = {}^t x^j D x^j$$
$$\text{cov}(x^j, x^{j'}) = D(x^j, x^{j'}) = {}^t x^j D x^{j'}$$

where: var = variance and cov = covariance.

c. Summary

The goal of PCA is to find subspaces $S(q) \in E$ with dimension q less or equal to p and preserving the maximum of information about data X. For $q = 1,...,p$:

$$\sum_{i=1}^{n} w_i \left\| P_{S(q)}(x_i) \right\|_M^2 = \underset{c \in S_q}{Max} \sum_{i=1}^{n} w_i \left\| P_c(x_i) \right\|_M^2$$

where S_q is the set of all the q-dimension subspaces of E and P_c is the orthogonal projection operator on the subspace $c \in S_q$.

d. Propriety

The subspace $S(q)$ is spanned by M-orthonormed eigenvectors $\{c_j\}$ of matrix ${}^t XDXM$ associated to the q first eigenvalues $\{\lambda_j\}$. These normed eigenvectors are called principal axis vectors and the eigenvalues principal moments. For all pair (λ_j, c_j) is associated a synthetic variable C^j called *principal component*. This variable is so that $C^j(i)$ is the co-ordinate of $P_{S(q)}(x_i)$ related to vector c_j. Two *principal component* are D-orthogonal.

2.2 Utility of principal elements

a. Partition of Variables

Working on multi-sensor fusion, we have associated the variables with the principal component which are closely correlated together , and it has leaded to their partition with a small cardinal while keeping the maximum of information, where groups of studied variables are substitued by principal component. The connections between these variables have been preserved (correlated and uncorrelated variables). The working data space has been reduced.

b. Data structure

In a space with dimensions up to 3, human eyes cannot distinguish the structures and the forms of the data. The PCA can manage to reduce data on principal subspaces with dimensions less than 3 and generally the principal plan given by the 2 first principal axes vectors. These representations are interpreted related to *principal component* associated to principal vectors considered. It results a easy interprstateion of group of studied variables associated to each *principal component*.

3. CLASSIFICATION METHOD

3.1 General Remarks

The classification consists in dividing subjects in homogeneous groups. Two subjects belonging to a same class are more similar than a subject belonging to a class and an other to a different class (Friedman, 1967; Cormack, 1971; Lerman, 1979).

Each class $P(k)$ is represented in the space E by its gravity centre defined by:

$$g(k) = \frac{\sum_{i \in P(k)} w_i x_i}{\sum_{i \in P(k)} w_i} = \frac{\sum_{i \in P(k)} w_i x_i}{w(k)}$$

The classification method used is the method which maximise the criteria of between-group dispersion, defined by:

$$B(P) = \sum_{k=1}^{q} w(k) \|g(k)\|_M^2$$

This method is equivalent to minimise the criteria of within-group dispersion, defined by:

$$W(P) = \sum_{k=1}^{q} \sum_{i \in P(k)}^{n} w_i \|x_i - g_k\|_M^2$$

Because, the total dispersion $T = \sum_{i=1}^{n} w_i \|x_i\|_M^2$ satisfy the following equation: $T = B(P) + W(P)$.

We know that Principal Component Analysis allows to group the variables. The classification method is similar, but here we have considered the subjects. Both the two methods are complementary and allow to consider either the variables or the subjects.

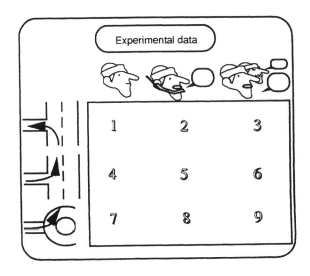

Fig. 1. The plot of the data base.

3.2 Representation

To visualise the structures revealed by a partition, we have projected the set of subjects on a plan spanned by two axial principal vector. We have obtained a easier interpretation related to variables strongly correlated to principal component associated with principal axis chosen.

4. APPLICATIONS

The previous analysis was applied to HUSAT experimental data. These data were obtained for three types of cross-roads (ONE, TWO, THREE) and for three driving conditions (CONTROL driving in silence, CARPHONE driving whilst conducting a conversation via a hand-free carphone, PASSENGER driving whilst conducting a conversation with a passenger), Fig. 1.

Our goal has been to distinguish the different cross-roads and driving styles. We have done a global analysis of the experimental data function of the distance *(d)* of the vehicle to the obstacle (cross-roads) to obtain a first idea of driver's behaviour. And then detailed analysis has been made to obtain more precise results. All the analysis has been done from 75 m to the obstacle (cross-roads).

4.1 Analysis

a. Related to the Cross-roads

For the CONTROL driver, we have obtained data from different cross-roads, that are precisely association of data 1, 4, 7 (Fig. 1.). Data for CARPHONE and PASSENGER drivers have been obtained by the same procedure ((2, 8) and (3, 6, 9) respectively). The correlation between initial

Driving condition		cp	variables				
			d	v	a	b	c
CONTROL	1, 4, 7	1	0.9887	0.9343	0.0000	-0.0354	-0.3576
		2	-0.1493	0.3533	0.0000	-0.2967	-0.6313
		:					
		5	0.0000	0.0000	0.0000	0.0000	0.0000
CARPHONE	2, 8	1	0.9814	0.9167	0.0000	-0.2360	0.0453
		2	-0.1920	0.3994	0.0000	0.3966	-0.2477
		:					
		5	0.0000	0.0000	0.0000	0.0000	0.0000
PASSENGER	3, 6, 9	1	0.9859	0.9293	0.0000	-0.5722	-0.0115
		2	-0.1670	0.3714	0.0000	0.2484	-3450
		:					
		5	0.0000	0.0000	0.0000	0.0000	0.0000

Table 1 The correlation between the variables and the principal component

The correlation between initial variables (d, v, a, b, c) and the *principal component* (1 ,..., 5) of different PCA are in Table 1.

The behaviour of each driver during the experiment is represented by principal plan (1,3) and (1,4) (Fig. 2.). The principal axis 1 is related to speed (v) and to the distance (d), the principal axis 3 is related to the clutch (c) and the principal axis 4 is related to the brake (b). We have observed that the driver's behaviours are different according to different cross-roads.

The 3 columns (CONTROL, CARPHONE, PASSENGER) of Fig. 2. represent the drivers, where: line is related to cross-roads ONE, big point is related to cross-roads TWO and small point is related to cross-roads THREE.

b. Related to the Driving Style

In order to have a good idea of different driving style related to a cross-roads. We elaborated files from the data obtained with different drivers related to a cross-roads. For example for cross-roads ONE, the file studied was obtained with 1, 2 and 3, (Fig. 1.).

The driver's behaviours are presented all along the trajectory. The 3 graphs (ONE, ONE, THREE) of Fig. 3. represent the cross-roads, where: line is related to driver CONTROL, big point is related to driver CARPHONE and small point is related to driver PASSENGER.

We observed that the CARPHONE driver used at least twice the clutch (Fig. 3a.). He used the brake before the other drivers (Fig. 3b.). The CONTROL driver used once the clutch all along the 74 m of the trajectory (Fig. 3a.). CARPHONE and PASSENGER drivers used the clutch on the cross-roads THREE (Fig. 3c.).

5. CONCLUSION AND PERSPECTIVES

Both the two methods we have described and applied to the data we have got have shown that they are well adapted to our vigilance detection problem. They distinguish the drivers having different behaviours all along their trajectory (or during the experiment) and to avoid lost of information in case of multisensor fusion.

Fig. 2. PCA related to the drivers.

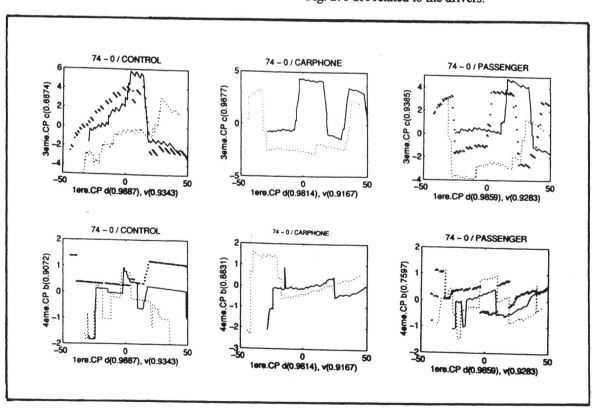

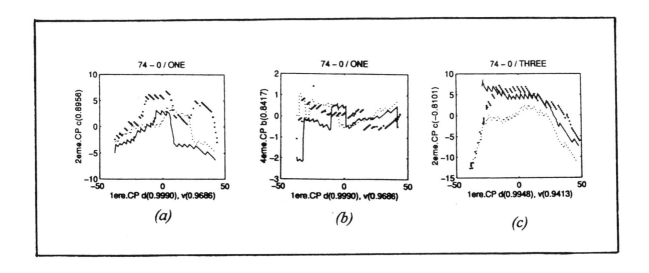

Fig. 3. PCA related to the cross-roads.

More over, both the methods can be implemented with artificial neural network. We are going to analyse some experimental data, these analysis allow us to have a knowledge base and classification of driver's behaviours all along their trajectory (during the experiment).

These analysis will allow us to characterise vigilance and no-vigilance driver states relatively to him behaviour evolution. So we can predict the state of a driver not considered, by comparing him evolution with its of states characterised below. In order to avoid accident, this prediction will be make from a certain distance of the obstacle.

REFERENCES

Benzecri, J. P., *Analyse des données*. Chapitre II B, Dunod, Paris 1973

Cailliez, F., J. P. Pages. *Introduction à l'analyse des données*. SMASH, Paris 1976

Celeux, G. et co-authors. *Classification automatique des données*. Dunod-Informatique, Paris 1989

Chan, M., Herrera, J. A., Martinez, D., Estève, D., André, B., Muzet, A., *Building and Testing of Expert System*. LAAS/CNRS, Strasbourg, France. Septembre 1992

Chan, M., Herrera, J. A., *Approche Neuronale pour la Détection Embarquée d'une Non-Vigilance du Conducteur Automobile*. Seminaire Vigilance et Transports: Aspects Fondamentaux, Dégradation et Prévention. Bron, France. decembre 1993

Cormack, M., *A review of classification*. Royal Journal Statistical Society, series A, 134, no. 3 pp. 321-367, 1971

Diday, D. et co-authors. *Optimisation en classification automatique*. INRIA, Rocquencourt, 1980

Friedman, H. P., Rubin, J., *One some invariant criteria for grouping data*. Journal of the Amaricain Statistique Association, no. 62, pp. 1159-1178, 1967

Herrera, J. A., *Evaluation de l'Etat de Vigilance du Conducteur Automobile par Réseaux de Neurones*. INSAT, Toulouse, France. Septembre 1992

Herrera, J. A., Chan, M., Estève, D., *Driving Safety with Neural Network*. 11th. European Annual Conference on Human Decision Making and Manual Control. Université de Valenciennes. Valenciennes, France. November 1992

Herrera, J. A., Labreche, S., Martinez, D., Chan, M., Estève, D., *Principal Component Analysis (PCA) and Classification: Application to Identify Driving Comportements*. Scientific Report LAAS/CNRS No. 94411, Toulouse, France. August 1994

Lerman, I. C., *Les présentations factorielles de la classification*. RAIRO, vol. 13, no. 2, pp. 107-128 and no. 33, pp. 227-251, 1979

Martinez, D. '*Offset, Une méthode de Construction Incrementale de Réseau de Neurones Multicouches et son Application à la Conception d'un Copilote Automobile*', Thèse de doctorat, Laboratoire d'Automatique et d'Analyse des Systèmes du CNRS, Toulouse, 1992

HIGH PRECISION AUTOPILOT DESIGN
FOR SMALL SHIPS

H. Loeb*, S. Ygorra* and M. Monsion*

This project is funded by the Port of Bordeaux Authority, the French National Agency for Innovation
(ANVAR) and the Regional Council of Aquitaine.

*Laboratoire d'Automatique et de Productique, Université de Bordeaux I, 351 cours de la Libération,
F-33405 Talence Cedex, FRANCE (E-mail: loeb@labri.u-bordeaux.fr).

Abstract. Ship autopilot technology has progressed considerably since the sixties. However most
existing autopilots are mainly heading control autopilots. The availability of new positioning systems
such as the Global Positioning System (GPS) and the possibility to obtain precise position thanks
to Differential GPS opens the way to a new generation of autopilots: track keeping autopilots which
would enable the ship to follow precisely a predefined path.
The Port of Bordeaux has undertaken with the help of the University of Bordeaux the development
of an autonomous surface vehicle (NAUSICAA) to serve as a platform for hydrographic data acquisition. This paper describes the general architecture of NAUSICAA and focuses on the design of the
control system. Results obtained in June 95 from actual trials on the Garonne River are presented.

Résumé. Les pilotes automatiques de bateau ont évolué considérablement depuis leur démarrage
dans les années soixantes. Mais ces pilotes automatiques restent pour l'instant des pilotes de maintien de cap. L'apparition de nouvelles techniques de positionnement tel que le Global Positioning
System (GPS) et la possibilité de travailler en différentiel (DGPS) ouvre la voie vers un nouveau type
de pilotes automatiques: les pilotes automatiques de maintien de route grâce auxquels un bateau
pourra suivre très précisément une trajectoire prédéterminée.
Le Port de Bordeaux a entrepris en collaboration avec l'Université de Bordeaux, le développement
d'un engin nautique de surface robotisé baptisé NAUSICAA. Cet article présente l'architecture
générale de NAUSICAA. Il décrit en particulier les techniques employées pour la synthèse du cor-
recteur. Les premiers résultats obtenus lors d'une campagne d'essais en juin 95 sur la Garonne sont
présentés.

Key Words. Track keeping, Ship Autopilot, DGPS, Optimal Control Theory.

1. INTRODUCTION

Ship autopilot design has already been a matter
of interest for many researchers. Most of the existing work was however done for big ships for obvious economical grounds. Many techniques have
been studied such as adaptive techniques, optimal control theory, and more recently H_∞ and
neural networks techniques. All of this work deals
mainly with heading control. A reference heading is entered in the autopilot. The controller is
responsible for making the necessary adjustments
to keep this heading. When navigating in open
seas, the outer loop (i.e., track keeping) is actually closed by hand with the helmsman defining
a new reference heading when necessary. Differential GPS techniques (based on satellites) now
enable the achievement of much more ambitious
objectives. DGPS positioning systems report the
position of a ship (or any vehicle) with a precision
in the meter range. Track keeping autopilots offer
therefore new possibilities for ship autopilots.

2. THE NAUSICAA PROJECT

The NAUSICAA project was initiated by the Hydrographic Service of the Port of Bordeaux. This
department is in charge of updating the maps necessary for navigation on the Garonne (the largest
estuary in Europe). Hydrographers are well aware
of the different positioning techniques available,
and their practical aspect. Navigation on hydrographic boats still requires a helmsman who follows the predefined trajectories on a computer
screen. The NAUSICAA project started as an
effort to increase productivity. It is a collaboration of the Port of Bordeaux with the University
of Bordeaux.

This project is ambitious. It consists in developing a small autonomous vehicle that would be a
convenient tool for hydrographers. NAUSICAA
is launched from a Mobile Control Station (MCS)
located on-shore or aboard a mother ship. Its
three modes of functionment suit different con-

figurations (safety problems, meteorological problems...):

- **Manual mode:** NAUSICAA is operated by a trained hydrographer on-board. The control panel furnishes the necessary information related to the mission. He can either run the ASV in the automatic mode and focus his attention on security and data acquisition, or pilot the ASV in a traditional way.
- **Teleoperated mode:** NAUSICAA is remotely controlled from the Mobile Control Station by the hydrographer.
- **Automatic mode:** The mission of the ASV is preprogrammed. It accurately follows the preregistered course. It is constantly tracked by the Mobile Control Station on-shore or aboard a mother ship. At any moment, the hydrographer can take over control in the teleoperated mode.

3. GENERAL FEATURES OF THE VEHICLE

NAUSICAA is unsinkable. It can be operated as far as 10 kilometers from the Mobile Control Station and has a survey autonomy of 4 to 5 hours. It can travel up to a speed of 10 knots to the mission site. The survey itself takes place at a speed of 4 to 8 knots. NAUSICAA relies on the following pieces of equipment:

- a pneumatic hull, roughly 5 meters long and 2 meters wide,
- a water jet propulsion unit, coupled to a 30 horsepower diesel engine,
- hydraulic drive units for control on the speed, direction and reverse gear,
- a fluxgate compass,
- a DGPS positioning system that delivers the information of position in the meter range and the information of speed over ground,
- an electromagnetic loch that delivers the information of speed over the water,
- a roll, pitch and heave compensator that delivers complete information concerning vehicle's attitude, and
- an echo-sounder system for measurement of depth below the boat.

4. OBJECTIVES OF THE TRACK KEEPING CONTROLLER

Track keeping is a classical problem in mobile robotics. Given a reference trajectory, the objective of the controller is to keep the vehicle on track, provided the disturbances do not exceed some limit. The reference track assigned to NAUSICAA consists simply of line segments. Further development will include circular arcs or other curves. The different types of trajectories actu-

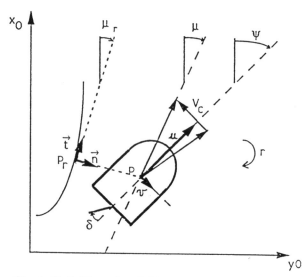

Fig. 1. Definition of variables

ally define the feedforward signal. Feedback signals are necessary to insure proper convergence of the error signals. The objective of the NAUSICAA track controller is to limit the cross track error distance from the reference trajectory to at most 5 meters.

The plant has here two inputs: control over ship speed and control over direction. Since there are few constraints concerning the ship speed, the system is decoupled. Speed regulation is therefore considered independently of direction regulation. It is therefore assumed that the ship is navigating at a constant longitudinal speed U which can be measured with the speed loch. Let P be the current position of the ship and P_r the projection of P on the reference trajectory. (See figure 4.) Two elementary errors can be defined by

$$\vec{P_r P} = e_n \vec{n}$$
$$e_h = \mu_r - \mu$$

where e_n is the cross track error and e_h is the heading error i.e., the angle formed by the tangent to the trajectory in P_r and the direction of speed over ground (Nelson and Cox, 1988). It is important here to contrast the definition of the ship's heading (as measured by a magnetic compass) with that of its speed heading (as measured by the DGPS). If the vehicle faces a strong transversal current, these two headings will be notably different. Another important point is to note that a vehicle such as NAUSICAA actually slides on the water during turns, and therefore maintains momentarily its previous trajectory. A sizable delay (4 seconds) is observed when comparing these two different headings (cf. figure 4).

A first order Nomoto model was used to design the controller. Identification tests were carried out in the Bordeaux wet dock so as to avoid currents

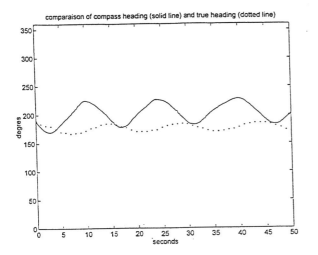

Fig. 2. Compared headings

(Chaumet-Lagrange *et al.*, 1994). The yaw rate signal r and rudder angle δ were recorded during the trial. The constants $K = 4s^{-1}, T = 6s$ varied moderately with the ship's speed U.

$$\frac{r}{\delta} = \frac{K}{1 + pT}.$$

5. CONTROLLER DESIGN: CLASSICAL APPROACH

The first approach to design the track keeping controller was derived from a classical approach. An inner loop was built to regulate the ship's heading using a PID controller. An outer loop was added to regulate position. In the outer loop only a PI controller has to be designed since the derivative action is performed by the inner PID loop.

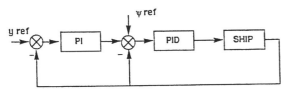

Fig. 3. Classical controller

Antiwindup techniques were used so as to limit the integral action. Simulations gave good results. NAUSICAA was then tested using this controller. Trials took place in Bordeaux. The results are shown on figure 5. It should be noted that for this trial the controller was fed with a cross track error and heading error that were calculated directly by the DGPS receiver. The cross track error was given as an integer number of meters. Disturbances during this trial (wind, current) were at a very low level.

Commercial autopilots use classically the cross track error and heading error that are calculated by the position receiver (NMEA standard). This

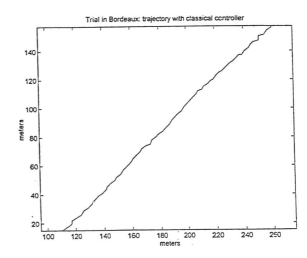

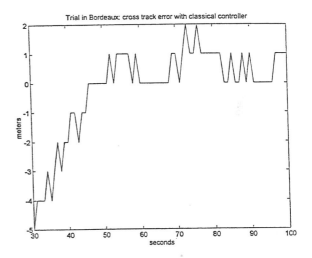

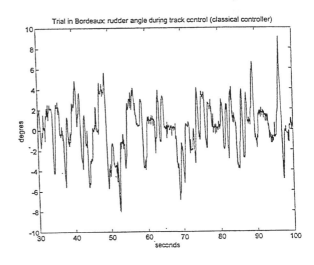

Fig. 4. Trial in Bordeaux with Classical Controller

is however a strong limitation for a track keeping controller because of the low update of the information. During turns no anticipation (feedforward signal) is possible. For good results in track keeping, the autopilot should have more information and be aware of the current segment to follow as well as the next segment (Holzhüter and Schulze, 1995).

6. CONTROLLER DESIGN OPTIMAL APPROACH

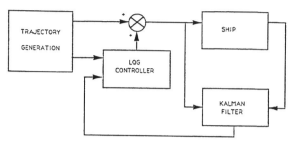

Fig. 5. Optimal controller

Optimal control theory was used to derive a track keeping controller (Loeb et al., 1995; Holzhüter, 1990). The control input δ is the sum of a feedforward control signal computed by the trajectory generator and the feedforward control signal computed using the LQG controller. The stochastic performance criterion J used for derivation of the controller gain uses the control input δ, heading error ψ, cross track error y and integral of cross track error z.

$$J = E(\lambda_\psi \psi^2 + \lambda_y y^2 + \lambda_z z^2 + \delta^2).$$

The notation used in mobile robotics (Nelson and Cox, 1988) (e_h for the heading error and e_n for the cross track error) is replaced here by the standard notation (ψ and y) used in marine applications (Fossen, 1994).

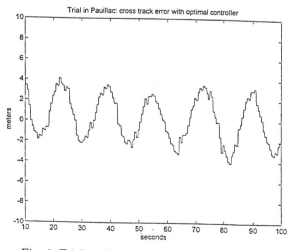

Fig. 6. Trial in Pauillac with Optimal Controller

Dynamics of the cross track error is expressed by

$$\dot{y} = U \sin(\psi)$$

which is linearized into

$$\dot{y} = U\psi$$

The state representation is then expressed by the following matricial equation

$$\begin{bmatrix} \dot{r} \\ \dot{\psi} \\ \dot{y} \\ \dot{r_b} \\ \dot{d_y} \end{bmatrix} = \begin{bmatrix} -1/T & 0 & 0 & 1/T & 0 \\ 1 & 0 & 0 & 0 & 0 \\ 0 & U & 0 & 0 & 1 \\ 0 & 0 & 0 & 0 & 0 \\ 0 & 0 & 0 & 0 & 0 \end{bmatrix} \begin{bmatrix} r \\ \psi \\ y \\ r_b \\ d_y \end{bmatrix}$$
$$+ \begin{bmatrix} K/T \\ 0 \\ 0 \\ 0 \\ 0 \end{bmatrix} \delta + \begin{bmatrix} 0 & 0 \\ 0 & 0 \\ 0 & 0 \\ 1 & 0 \\ 0 & 1 \end{bmatrix} \begin{bmatrix} w_1 \\ w_2 \end{bmatrix}$$

where r_b is the torque applied to the ship, d_y is a transversal current, and w_1 and w_2 are two white noises.

The reduced "*" system is used to derive a controller whose dynamics are independent of the ship's speed U. A reduced equivalent criterion

$$J^* = E \left(\lambda_\psi^* (\psi^*)^2 + \lambda_y^* (y^*)^2 + \lambda_z^* (z^*)^2 + (\delta^*)^2 \right)$$

is used with reduced variables

$$
\begin{aligned}
t^* &= t/T & \delta^* &= \delta KT \\
r^* &= rT & \psi^* &= \psi \\
y^* &= y/(UT) & z^* &= z/(UT^2)
\end{aligned}
$$

$$
\begin{aligned}
\lambda_\psi^* &= (KT)^2 \lambda_\psi \\
\lambda_y^* &= (UT^2)(KT)^2 \lambda_y \\
\lambda_z^* &= (UT^2)^2 (KT)^2 \lambda_z.
\end{aligned}
$$

Let (p_{ij}) be the solution of the Ricatti equation. The controller output, i.e., rudder angle, can then be computed by

$$\delta = - \left(p_{11} \hat{r} + p_{12} \hat{\psi} + p_{13} \hat{y} + p_{14} \hat{z} \right)$$

where $\hat{r}$, $\hat{\psi}$, $\hat{y}$ and $\hat{z}$ are the estimates of the yaw rate, heading error, cross track error, and integral of cross track error respectively as computed by the Kalman filter.

The simulations run gave satisfactory results. Next, NAUSICAA was tested in Paullac on the Garonne river (cf. figure 6). Cross track error stayed almost constantly below 5 meters. A low frequency oscillation (0.35 radians) was however observed. That can be explained by the use of the raw heading angle in the feedback law. Further development will include the design of a Kalman filter to filter out measurement noise. An estimate for the transversal current will enable us to define a new reference heading for the ship's heading.

Fig. 7. View of Prototype

7. PRACTICAL ASPECTS OF DEVELOPMENT

An embedded computer was installed in NAUSICAA, with both analog and numerical data acquisition cards. The control system was developed using C language. Heading information is continuously available, position information (from the DGPS receiver) is updated every 0.6 seconds. The control procedure is processed every 0.1 seconds. Limitations on the performances are due to the rudder speed which impose a limit on the controller's bandwidth

$$\omega < \frac{\pi v}{2A}$$

where v is the rudder speed, and A is the largest amplitude of the rudder. For NAUSICAA, $v = 12$ deg/s and $A = 30$ deg so that $\omega_{max} = 0.62$ rad/s.

Further tests will be carried out in September 95 to study the robustness of the track controller. A comparative study of the classical and optimal controller will be done.

8. REFERENCES

Chaumet-Lagrange, M., H. Loeb and S. Ygorra (1994). Conception d'un engin nautique de surface robotisé (E.N.S.R.). In: *Proceedings OCEANS '94*. Vol. I. Brest, France. pp. 120–129.

Fossen, Thor I. (1994). *Guidance and Control of Ocean Vehicules*. John Wiley and Sons Ltd.. Chichester, England.

Holzhüter, T. (1990). A high precision track controller for ships. In: *Proceedings of IFAC 11th Triennial World Congress*. Talinn, Estonia, USSR. pp. 425–430.

Holzhüter, T. and R. Schulze (1995). Operating experience with a high precision track controller for commercial ships. In: *Proceedings of the 3rd IFAC Workshop on Control Applications in Marine Systems*. CAMS'96. Trondheim, Norway. pp. 270–277.

Loeb, H., S. Ygorra and M. Monsion (1995). New hydrographic automated vehicle: Design of a high precision track controller. In: *Proceedings of the 3rd IFAC Workshop on Control Applications in Marine Systems*. CAMS'96. Trondheim, Norway. pp. 49–53.

Nelson, W.L. and I.J. Cox (1988). Local path control of an autonomous vehicle. In: *Proceedings of the 1988 IEEE Conference on Robotics and Automation*. Philadelphia, Pennsylvania, USA. pp. 1504–1510.

Witt, N. A. J., R. Sutton and Miller K. M. (1995). A track keeping neural network controller for ship guidance. In: *Proceedings of the 3rd IFAC Workshop on Control Applications in Marine Systems*. CAMS'96. Trondheim, Norway. pp. 385–392.

NEW RULES FOR OPTIMIZING FUEL SAVINGS FOR AGRICULTURAL TRACTOR

O. Muller, O. Naud, C. Cedra

*Cemagref, Electronic and Artificial Intelligence Laboratory,
Parc de tourvoie, BP 121, 92185 ANTONY Cedex, France*

Abstract: The Cemagref is working on a dashboard display providing information about tractor efficiency. The goal is to encourage more performant and environmental-friendly tractor use. The operator is helped to tune the system composed by the tractor and the attached implement dynamically during work. A spot is developed about saving fuel by the display of appropriate gear and throttle position for current work. The "Gear up and throttle down" method is used. A new knowledge representation and new processing rules are presented.

Keywords: agricultural tractor, motor, gear box, power, advice, fuel consumption, adjustment.

1. INTRODUCTION

During agricultural work, the operator needs to tune the tractor-implement system when the engine load changes (caused by soil variations): he must change the engine speed and the gear setting to maintain the same working speed, and to preserve power and torque reserve.

The Cemagref is working on the display of information onboard the tractor. One purpose of the research is to tell the operator which appropriate gear and throttle position to choose for a given engine power.

The "Gear up and throttle down" method (noted GUTD, i.e. to select an higher gear and to reduce the engine speed) has been adapted so that the operator can save fuel while keeping driving flexibility during work.

The energy savings result from better combustion, reduced internal engine friction and lighter loads from the engine fan, hydraulic pump, and other accessories.

These potential fuel savings depend on a number of factors, including power level, fields variability, engine torque characteristics, and constraints when operating PTO (Power Take-Off) driven equipment (Stephen *et al.* , 1981).

The GUTD process requires a mean of describing where a tractor engine is operating on its performance map.

A new method and associated data processing are proposed to determine the appropriate gear and engine speed according to a given engine power. A new graphical representation and a new advice form are also proposed for a powerful visual control onboard the working tractor .

2. LITERATURE SURVEY

Different systems have been developed for supplying current performance information to the operator. Some systems merely supply information, some provide instructions for improving efficiency, and a few automatic control systems have been developed for optimizing machine performance.

Schrok *et al.* (1982) worked about a system indicating when "Shift up throttle back" is feasible : it consists in calculating the engine speed and torque required to maintain a constant forward speed in the next higher gear. The system checks that the resultant torque is less than 80 % of the maximum torque available at the engine speed, using the performance map model developed by Jahns (1983). A limit of 80 % was imposed to provide a torque reserve. The previous step is repeated until either the highest gear was reached or the torque exceeded the 80 % limit.

The more frequently checks are made, the more frequently messages can be displayed. The operator is told what the new engine and gear setting should be. The operator is encouraged to make the change by being told what fuel savings should result from this new setting. The message remains visible on the terminal for the duration of the time period between feedback checks, giving the operator ample time to observe the new setting.

Chancelloor and Thai (1983) developed a hard wired control system which used the two input signals, torque sensor and forward desired speed, and which varied both transmission ratio and engine speed.

Grogan *et al.* (1984) utilised Schrock's procedure and worked about this system. He used a voice synthesiser to communicate with the operator (to augment visual displays of messages with corresponding verbal messages). This system was also able to predict the fuel consumption for the highest allowable gear and compare this with the current measured fuel consumption.

Wang and Zoerb (1985) developed a monitor which gave a zero, positive or negative output when the engine ran at the ideal combination or not.

3. LAST NOVELTIES FROM MANUFACTURERS

Engine speed is easily measured but power has to be measured through indirect values. These indirect values are exhaust gas temperature ("Acet" system; Renault tractor) and fuel injection time measured at the injection nozzle ("Infomat" system; Steyr tractor), or is now indirectly given by the electronic injection pump.
The last system from Steyr not only advises the operator if it should be appropriate to select the next higher or lower gear, but is also able to engage the next right gear automatically since the tractor has an automatic gear box. The driver is only informed which gear is engaged. Renault and Steyr are giving up these devices because it seems that they do not comply with the customers.
The powershift gear box allows the operator to shift gear without action on the clutch pedal. Since it appeared (a few years ago on tractor), some recent improvements have been realised : nowadays, you can work on a tractor with a semi automatic gear box (e.g. Same, which won this year a distinction at last SIMA (Paris International Agribusiness Show), and Ford (functionality available for a theoretical speed > 11 km/h). Notice that even if gears automatically change, engine speed do not vary a lot because of the action of the governor.

4. NEW NEEDS

Today the gear number has considerably increased. It is not rare to overstep forward 30 gears with the new power shift gears (e.g. Deutz with its "Agro star"

which allows 48 forward gears, Fendt with its "Favorit " 40 forward gears). Gear multiplicity will involve a higher combination than before, that will allow to work nearer 80 % maximum engine torque for a determined engine speed. The necessity for an advice or a help about which gear and throttle position to choose is more relevant than ever. In this situation, it should be difficult to use Schrock method since the result could be a never endless gear shifting for the operator.

A first alternative is the development of a full automatic powershift box with an electronic management of the injection pump. The operator should not be incited to do himself Shift Up and Throttle Down continuously and his task could be to inform the system about his wondered forward speed. But in this case a full automatic gear box should work continuously

A second alternative is to select immediately the right gear and the right throttle position (instead of the step by step procedure proposed by Shrock and Grogan), stay below enough full load, and determine good working *areas* for the engine (instead of a straight advice line on the engine map).

This second approach is submitted here below.

5. OUR CONCEPT

5.1 A calculus method of the gear and throttle position

A calculus method is proposed to determine an appropriate gear and throttle position according to the engine load and the different gears.
On the map of your engine, the first difficulty is to draw the advice curve giving the relationship between suitable engine speeds and engine loads (see farther engine map, figure 1) Then an onboard electronical device can use the following procedure when the tractor is working:

* On the same map, for the current engine load and engine speed, check if the resulting point is, or not, situated in the white area.
* If not, select and store the advised engine speed according to the measured load, on the advise dotted curve (see engine map, figure1),
* Get now your current theoretical speed,
* Let us divide and find X1 as :

$$X1 = \frac{\text{Theorical speed (km/h)}}{\text{Engine Speed (rpm)}}$$

and store it,
* For each X of the following board, table1,
    Subtract X1 (stored) to X
* Choose the minus result (absolute value), select the corresponding new gear ratio and its appropriate X, and store them,
* Select the theoretical travelling speed and multiply it with the X selected (the result gives the new appropriate engine speed for the same travel speed).

Table 1 kinematic parameters

Range	Gear	v*= Fwd speed (ω =2350 rpm) km/h	$X = \dfrac{v^*}{2350\,\text{rpm}}$
1	1 st Slow	2,16	0,0009191
1	1 st Fast	2,67	0,0011362
1	2 nd Slow	3,43	0,0014596
1	2 nd Fast	4,24	0,0018043
1	3 rd Slow	5,38	0,0022894
1	3 rd Fast	6,66	0,002834
1	4 th Slow	8,48	0,0036085
1	4 th Fast	10,50	0,0044681
2	1 st Slow	6,44	0,0027404
2	1 st Fast	7,98	0,0033957
2	2 nd Slow	10,25	0,0043617
2	2 nd Fast	12,69	0,0054
2	3 rd Slow	16,09	0,0068468
2	3 rd Fast	19,93	0,0084809
2	4 th Slow	25,37	0,0107957
2	4 th Fast	31,42	0,0134979

v* specified for back tyre type 20.8.38
ω = engine speed

Example : A current working situation (point A on the map) is defined by the following parameters :
* Engine developed power :   62 kW
* Theoretical speed :   8,8 km/h
* Engine speed :   1625 r.p.m.

For the same travel speed, the advised situation should be the one defined by the point B on the map, since it allows a better power and torque reserve. There is a better use of the engine and improvement driving comfort, even if the specific consumption increases, and fuel consumption too. The advised situation is the following one :
* Engine developed power : 62 kW
* Theoretical speed :   8,8 Km / h
* Engine speed (new) :   1950 r.p.m.

The calculus of X1 gives : X1 = (8,8/1950) = 0,0045182; the nearest X is 0,0044681 and the corresponding gear is "1 range 4th Fast". To keep with exactly the same travel speed (8,8 km/h), the engine speed has to be slightly modified. The appropriate engine speed will be :
8,8/0,0044681 = 1970 r.p.m. (which is a little different from the initial 1950 r.p.m.).

Imagine another current working situation (point C on the map); for the same travel speed, the advised situation should be the one defined by the point D on the map, since it allows a fuel saving. This advised situation is the opposite to the previous one, but in this case, a better specific consumption and sufficient power reserve (the potential action of the governor is very high) are obtained.

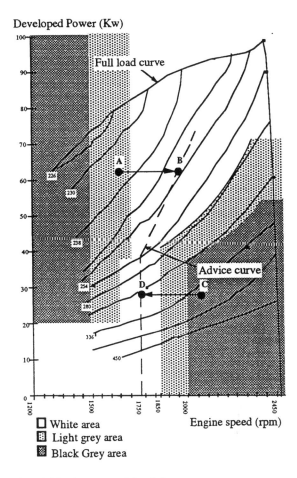

Fig. 1. Engine map with advice curve

5.2 Graphical presentation of the optimisation driving areas

The engine map can be displayed on the onboard screen. Thus, the operator is able to notice where the engine is working (illustrated by a luminous point moving on the map).

Three different coloured areas are drawn, meaning correct use or not :
- good engine use
- medium use, that could be improved (light grey)
- wrong use, which leads to a visual warning on the onboard screen (grey black).

* Black grey zone (see engine map, Fig. 1) deals with a *really wrong* use of the engine: there is a winking red light around the engine picture on the principal window of the dashboard. A warning message is displayed, incitating the operator to look on the advise message for a new gear and throttle position.
* Light grey zone deals with a medium use deals with a *light* wrong use of the engine: there is a winking orange light around the engine picture. You can have a look on the advise message, as before, but without warning message.
* White zone means a good functioning situation of the engine.

121

CONCLUSION

One of the tractor trends is to develop gear multiplicity that will involve a higher combination than before. The purpose is to tell the operator which appropriate gear and throttle position to choose for a given engine power by using the "Gear up and throttle down" method.

Good working *areas* are determined, taking into account at the same moment fuel saving, ergonomic care (the operator should not be incited to Shift Up and Throttle Down continuously) and engine load changes, caused by soil variations (that means that a power and torque reserve is necessary). It is a compromise to advise either fuel saving or driving comfort being sure to get a sufficient torque reserve.
A procedure to select immediately the right gear and the right throttle position is proposed and also a new way to represent the working state of the engine, giving the operator fully understanding of the advice basis.

At the present time at the Cemagref centre, an advice GUTD system equips a mechanical gear box and the operator shifts himself the gear handles and the throttle handle. Beyond this first improvement, the real innovation is coming from the fact that one might couple this device with a "power shift" gear box. It would be a great progress, allowing an "automatic adjustments tractor": the tractor might preserve automatically the same travelling speed, whatever the variations of engine power required by the tillage operations are.

But this is true for works without PTO which requires always the same engine speed. With a PTO work, the advice procedure is nearly the same, except the engine speed which should be the same

REFERENCES

Chancelloor, W.J., N.C.Thai (1983). Automatic Control of tractor transmission ratio and engine speed. *ASAE, paper N° 83-1061,* pp 642-646.

Grogan, J., D.A. Morris, S.W. Searcy, H . Wiedemann, B.A. Stout (1984). Micro computerbased information feedback system for improving tractor efficiency. *ASAE Paper N° 84-1594,* 13 p.

Jahns, G. (1983). A method of describing diesel engine performance maps. *ASAE Paper N° NCR 83-101.*

Jahns, G., H. Speckmann (1988). Driver aids for optimal tractor utilization. *ASAE N° 88-1062,* 8 p.

Morris, D.A., S.W. Searcy, B.A. Stout (1987). On board tractor Microcomputer system. *Agric. Engng Res.* **38.**

Wang G. and G.C. Zoerb (1985). A tractor gear selection; indicator. *ASAE Paper N° 85-1051,* 14p.

Stephen L.E., A.D. Spencer, V.G. Floyed, W.W. Brixius (1981). Energy requirements for tillage and planting. *ASAE Pub. 4-1981 (American Society of Agricultural Engineers).*

Schrock, M.D., D.K. Matteson, J.G. Thompson (1982). A gear selection aid for agricultural tractors. *ASAE paper N° 82-5515,* 15 p.

MODELING AND CRONE CONTROL OF A VERY HANDY MOBILE BASE

Hervé Linarès, Jocelyn Sabatier, Michel Nouillant and Alain Oustaloup

CRONE team, Laboratoire d'Automatique et de Productique - ENSERB - Université Bordeaux I
351, cours de la Libération - 33405 TALENCE Cedex - FRANCE
Tel. (33) 56 84 61 40 - Fax. (33) 56 84 66 44
E-mail : oustalou@lap.u-bordeaux.fr

Abstract : This paper presents the modeling, the linearisation and the third generation CRONE control of a mobile robot fitted with a base with four drive and steering wheels. The non-linear dynamic model takes into account the tire slips, the hydrostatic transmission features, the differentials and the free pivot on the front-wheel-axle unit. A modification of the state vector of the non-linear system makes it possible through a first order linearisation, to obtain a set of linear plants representative of the non-linear plant for non-zero speeds. CRONE regulators ensuring the robust trajectory tracking are then synthesized. The robustness of the CRONE control enables the mobile base to be brought back to the planned trajectory with a robust degree of stability, despite parametric variations of the base and ground features.

Keywords : dynamic modeling, first order linearisation, CRONE control, mobile robot.

1. INTRODUCTION

The purpose of the DACTARIX robot ("Dispositif Autonome Conçu pour Travaux A RIsques sur eXplosifs",or autonomous device for dangerous work with explosives) designed and developed by the C.E.A. consists of picking up projectiles containing shocked explosives. As a result, it has to move in an environment with variable parameters. Studies on stability of the control of non-holonomic robots have already been performed (Kanayama, *et al.*, 1991; Canudas de Wit, *et al.*, 1992). It is nonetheless important to take into account the problems of robustness. In this field, the controls developed often use simplified models (Ackermann, *et al.*, 1993). In this article, after a description and complete modeling of the mobile base presented in sections 2 and 3, the synthesis method of the CRONE regulators providing the robust trajectory tracking is detailed in section 4. This synthesis is based on a set of linear plants representative of the non-linear plant for all the variations of the parameters considered.

Such plants are obtained by first order linearisation of the non-linear model, after a modification of the state vector which makes it possible to take account of the speed in forward drive. The control structure adopted makes it possible to cancel out errors of orientation and position of the base. The simulation of the corresponding control law is presented in section 5. The robustness of the CRONE control enables the mobile base to be brought back to the planned trajectory with robust dynamics of damping, despite variations in the parameters of the base and of the ground features.

2. DESCRIPTION OF THE MOBILE BASE

The mobile base (fig. 1) of the DACTARIX robot has four drive and steering wheels. This base, of simple design, has no suspension, but a good stability provided by a horizontal pivot in the middle of the front axle unit. Drive power is provided by a centrifugally regulated diesel engine.

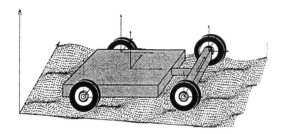

Fig. 1. Mobile base

The hydrostatic transmission provides easy control of the forward drive of the base without any drop in torque at low speeds. Power is then transmitted to the wheels through two differentials. The simplicity of the steering device imposes the same angle of lock on both the right-hand and left-hand wheels. This constraint fails to observe Jeantaud's diagram, and subjects the low pressure tires to lateral stresses, resulting in the appearance of slip that must be tacken into account in modeling.

3. MODELING

The dynamic equations describing the behavior of the mobile base can be represented by the loop system shown in figure 2. The forces likely to cause evolution of the dynamics of the base are those applied by the ground on the wheels. They consists of the reaction to the weight of the vehicle, the reaction to the forward drive torque, and the reaction to the side slip (fig. 3). They are calculated by taking into account the geometric and dynamic features of the various components of the base, and the coefficient of sinking into the ground and the tire adherence of the ground.

The longitudinal slip is defined by :

$$g = \frac{R_{wheel}\omega_{wheel} - V_{realwheel}}{R_{wheel}\omega_{wheel}} . \qquad (1)$$

It depends on the load on the wheel, on the quality of the tire/ground adherence and on the torque transmitted (Wulfsohn, et al., 1992). A function of the same type links the angle of side slip to the lateral force on the tire. The instantaneous load on the wheel is calculated from the flattening of the tire, given that its stiffness and structural damping are known for a given inflation. The differentials are modelized by torque scales.

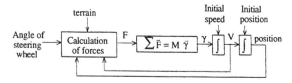

Fig 2. Structure of the dynamic model

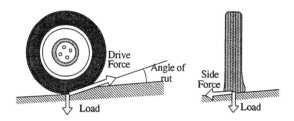

Fig 3. Modeling of the Wheel/Ground contact

The ground reaction to the three forces thus defined depends on its orientation in relation to the wheel. The ground is therefore modelized locally by the plane tangential to it at the point of contact of the wheel. The reaction forces can then be calculated and brought back to the mass center of the base. At this point, its accelerometric behavior can be deducted from the torquer of the forces and torque according to the three axes. The integration of acceleration gives the speed and then the position of the base at any given point in time. The only input enabling the control of the mobile base is the angle of lock. The model of the mobile base thus defined is shown in figure 4.

4. CONTROL

The behavior of the mobile base must have low sensibility to modifications of its forward drive speed and of its environment. The synthesis method of the control laws ensuring tracking of the trajectory must therefore take account of these problems of sensibility. In the case of the CRONE control (Oustaloup, 1991), this is made possible by considering a set of linear plants representative of the non-linear dynamic behavior of the base for different forward drive speeds and for different environmental parameters.

4.1 Linearisation

The linear plants thus defined are obtained by determining the first order approximation (tangent linear model) of the non-linear model. This linearisation is based on a Taylor series expansion of the non-linear model at an equilibrium point (a point where the derivative of the state vector is nil) (Fossard, et al., 1993). In the case of the mobile base, as its position is part of the state vector of the non-linear model, all the equilibrium points correspond to a state of the base where the speed (derivative of the position) is nil. The tangent linear models namely for forward drive speeds near zero. The solution to this problem was obtained by transforming the non-linear model so as to eliminate the position of the base from the state vector.

124

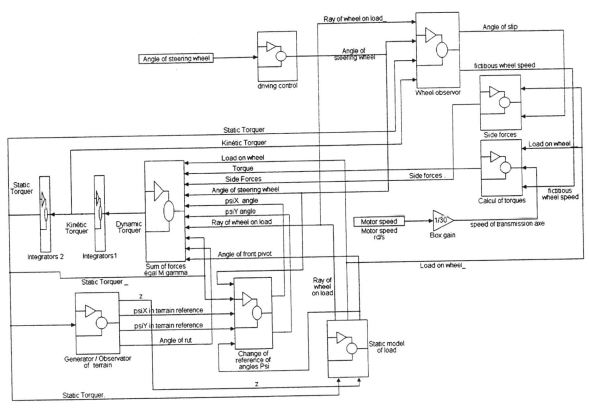

Fig. 4. Diagram of non-linear model of the base

The dynamical equation of the non-linear model is written :

$$\begin{cases} \dot{X} = g(X,U) \\ Y = h(X,U) \end{cases}, \tag{2}$$

with

$$\begin{cases} X = \begin{bmatrix} x_g\ y_g\ z_g\ \theta_r\ \theta_t\ \theta_l \end{bmatrix}^T \\ Y = \begin{bmatrix} \varepsilon_{pos}\ \varepsilon_{ori} \end{bmatrix}^T \quad \text{and} \quad U = \begin{bmatrix} \theta_{vol} \end{bmatrix} \end{cases}, \tag{3}$$

where

$(x_g\ y_g\ z_g)$ are the coordinates of the mass center,
$(\theta_r\ \theta_t\ \theta_l)$ are the angles of orientation in roll, pitch and yaw,
θ_{vol} is the angle of the steering wheel and ε_{pos} and ε_{ori} the differences in position and orientation.

By considering that the mobile base moves at a constant speed, namely $V = [V_x\ V_y\ V_z\ \Omega_r\ \Omega_t\ \Omega_l]^T$, the derivative of the state vector can be written as

$$\begin{aligned} \dot{X} &= \begin{bmatrix} V_x + \varepsilon_{vx}\ V_y + \varepsilon_{vy}\ V_z + \varepsilon_{vz}\ \Omega_r + \varepsilon_{\Omega r}\ \Omega_t + \varepsilon_{\Omega t}\ \Omega_l + \varepsilon_{\Omega l} \end{bmatrix}^T \\ &= \begin{bmatrix} V_x\ V_y\ V_z\ \Omega_r\ \Omega_t\ \Omega_l \end{bmatrix}^T + \begin{bmatrix} \varepsilon_{vx}\ \varepsilon_{vy}\ \varepsilon_{vz}\ \varepsilon_{\Omega r}\ \varepsilon_{\Omega t}\ \varepsilon_{\Omega l} \end{bmatrix}^T, \\ &= V + \dot{\tilde{X}} \end{aligned} \tag{4}$$

where ε correspond to small variations.

The base can be modelized by a non-linear system of the form :

$$\begin{cases} \dot{\tilde{X}} = \tilde{g}(\tilde{X},U) \\ Y = \tilde{h}(\tilde{X},U) \end{cases}, \tag{5}$$

with

$$\begin{cases} \tilde{g}(\tilde{X},U) = g(X,U) - V \\ \text{and } \tilde{h}(\tilde{X},U) = h(X,U) \end{cases}. \tag{6}$$

The equilibrium of this new system corresponds to a movement of the base at constant speed. Linearisation by first order approximation of the system (5) now makes it possible to obtain a linear system representative of the non-linear system (2) for non-zero speeds. By varying the parameters of the system (5), a set of transfer matrices G(p) reflecting the dynamic behavior of the non-linear system for the variations considered is created by linearisation. From the set of G(p) tranfers thus obtained, a nominal transfer is taken, $G_0(p)$, and a set of uncertainties domains is created (Lanusse, *et al.*, 1993), significant of the gain and of the phase of these tranfers at each frequency.

4.2 The third generation CRONE control

The objectives of the CRONE control (*Commande Robuste d'Ordre Non Entier*) is to minimize the variations of the stability degree of the control loop in spite of plant uncertainties (structured or not). Three strategies make it possible to reach this objective.

125

The first strategy (first generation CRONE control), consist in reducing the phase margin variations of the control, to the phase variations of the plant around the open loop unit gain frequency ω_u. Around ω_u, the regulator in cascade with the plant, is defined by a non integer order transmittance, namely :

$$C(s) = \left(\frac{s}{\omega_0}\right)^n \qquad \text{with} \quad n \in R . \qquad (7)$$

The second strategy (second generation CRONE control), more demanding than the first one, consists in cancelling the phase margin variations of the open loop directly by securing two conditions :
- a vertical template formed by the open loop Nichols locus for the nominal parametric state of the plant ;
- a vertical sliding of the template when a reparametration of the plant is performed. The template thus defined is described by the transmittance of a real non integer integrator whose order determines its phase placement, namely :

$$\beta(s) = \left(\frac{\omega_u}{s}\right)^n \qquad \text{with} \quad 1 \le n \le 2 . \qquad (8)$$

When the second condition cannot be secured, there is no reason why the vertical template should be the one that achieves the best robustness of the control. One should thus consider a template, always defined as a straight line segment for the nominal parametric state of the plant, but of indifferent direction, called "*generalized template*" (fig. 5).

There exists an indefinite number of generalized templates which can tangent the same iso-overshoot contour (of graduation Q).

Among the infinity of generalized templates thus defined, it is convenient to select an *optimal template* in conformity with the minimization of a quadratic criterion at the time of a reparametration of the plant.

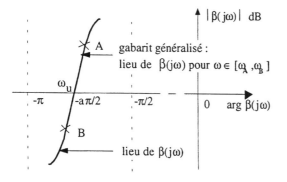

Fig. 5. Representation of the generalized template by an indifferent direction straight line segment in the Nichols plane

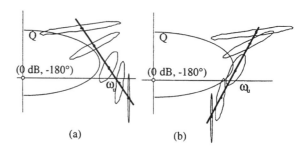

Fig. 6. The optimal approach allowed by the generalized template ensures an optimal placement of the uncertainty domains : (a) indifferent generalized template ; (b) optimal template

By minimizing such a criterion, the optimal template positions the uncertainty domains (Lanusse, *et al.*, 1993) correctly, so that they overlap as little as possible on the low stability degree areas (fig. 6). Trying to synthesize such a template defines the initial approach of the third generation CRONE control.

The generalized template can be characterized by a complex non integer order of integration, n, whose the integer part, a, determines the phase placement at frequency ω_u, and then, whose imaginary part, b, then determines its incline in relation to the vertical (fig. 5).

The generalized template is described by a transmittance based on the transmittance of a complex non integer integrator (Lanusse, *et al.*, 1993), namely :

$$\beta(s) = y_0 \left(ch\left(b\frac{\pi}{2}\right)\right)^{\text{sign}(b)} \left(\frac{\omega_r}{s}\right)^a \left(cos\left(b \ln\left(\frac{\omega_r}{s}\right)\right)\right)^{-\text{sign}(b)} \qquad (9)$$

for $\omega \in [\omega_A, \omega_B]$, $n = a+ib$.

The aim is an analytical description of the behavior in open loop for the nominal plant, which would take into account at the same time :
- the plant behavior in the low frequencies, to ensure a good steady state accuracy ;
- the generalized template around unit gain frequency ω_u ;
- the plant behavior in the high frequencies.

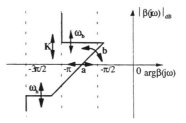

Fig. 7. Visualisation of the effects of the five degrees of freedom on $\beta(s)$

It turns out that the behavior thus defined can be described by a transmittance based on the *frequency limited complex non integer integration* (Lanusse, *et al.*, 1993), namely :

$$\beta(s) = K \left(\frac{\omega_b}{s} + 1\right)^{n_b} \left(\frac{1 + \dfrac{s}{\omega_h}}{1 + \dfrac{s}{\omega_b}}\right)^a$$

$$\left(\mathcal{R}e\left[\left[C_0 \frac{1 + \dfrac{s}{\omega_h}}{1 + \dfrac{s}{\omega_b}}\right]^{ib}\right]\right)^{-sign(b)} \frac{1}{\left(1 + \dfrac{s}{\omega_h}\right)^{n_h}} \ , \quad (10)$$

where

$$C_0 = \left[\left(1 + \frac{\omega_r^2}{\omega_b^2}\right) \middle/ \left(1 + \frac{\omega_r^2}{\omega_h^2}\right)\right]^{1/2} \quad . \quad (11)$$

K ensures the open loop unit gain frequency ω_u set by the designer. ω_b and ω_h define the transitional frequency. n_b and n_h are respectively the asymptotic behavior orders in open loop in low ($\omega < \omega_b$) and hight ($\omega > \omega_h$) frequency. a and b are the real and imaginary orders of integration. ω_r is the resonance frequency close to ω_u.

The optimisation of the generalized template regarding
- its position (along the 0 dB axis)
- its direction (or incline in relation to the vertical)
- its length and its prolongation in low and high frequencies,
consists in determining the five optimal parameters of the nominal open loop transmittance $\beta(s)$:
- optimal real integration order, a_{opt}, and optimal gain, K_{opt}
- optimal imaginary integration order, b_{opt}
- optimal transitional frequencies, ω_{bopt} and ω_{hopt}.

Figure 7 shows how these parameters are instrumental in the search for the optimal open loop transmittance $\beta(s)$.

The unit gain frequency and the tangency to an iso-overshoot contour being set, only three independent parameters are to be considered. So the search for the optimal parameters is simplified.

The C(s) optimal regulator in cascade with the plant is synthetized from its frequency response deduced from relation

$$C(j\omega) = \frac{\beta(j\omega)}{G_0(j\omega)} \ . \quad (12)$$

4.3 Synthesis of the CRONE regulators

The control law which will be implemented on the mobile base must ensure proper trajectory tracking for all the plants considered in section 4.1. The control diagram envisaged is that shown in figure 8.

This control scheme includes a fast dynamic orientation servo-control, the input of which is given by a much slower dynamic position regulator. The regulators C_{ori} and C_{pos} are synthesized according to the third generation CRONE approach. When the position error is nil, this structure makes it possible to cancel out the orientation error of the base, in particular that caused by slippage.

Given the difference of dynamic between the loops, for the first one, the plant is defined by the transmittance $G_{ori}(s) = \varepsilon_{ori}(s) / u(s)$. The plant used for the second loop is $G_{pos}(s) = C_{ori}(s)G_p(s) / (1+C_{ori}(s)G_{ori}(s))$, where $G_p(s) = \varepsilon_{pos}(s) / u(s)$.

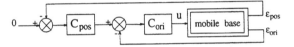

Fig. 8. Control diagram of the mobile base

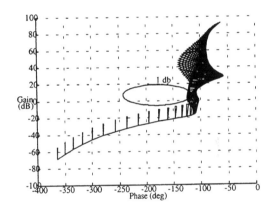

Fig. 9. Nichols locus of $C_{ori}(j\omega)G_{ori}(j\omega)$ and associated uncertainty domains

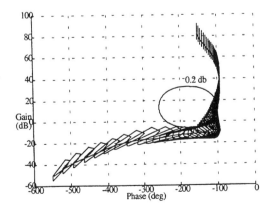

Fig. 10. Nichols locus of $C_{pos}(j\omega)G_{pos}(j\omega)$ and associated uncertainty domains

As illustrated by the Nichols locus of the two open loops $C_{ori}(s)G_{ori}(s)$ and $C_{pos}(s)G_{pos}(s)$ (fig. 9 and 10), the third generation CRONE control ensures the robustness of the stability degree by moving the uncertainty domains of plants so that they overlap as little as possible on the low stability degree areas.

Such a property is obtained by minimizing the most extreme variations of the closed loops resonance ratios in tracking.

5. PERFORMANCES

To valide the strategy developed in the preceding section, the control shown in figure 8 (including the non-linear model (2) of the base) was subjected to a step test. For the nominal plant and for the maximum and minimum variations in weight, speed, ground adherence and sinking coefficient, the base was launched on a trajectory forming a step of 1 m.

As illustrated in figures 11 and 12, the control law make it possible to obtain a high robustness of the first overshoot of the position response with respect to the variation in parameters, and a good rapidity in compliance with the mechanical limits of the base.

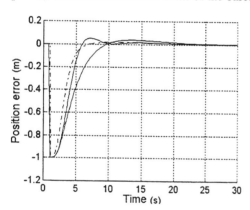

Fig. 11. Error in position
··· nominal plant ; — extremal plants

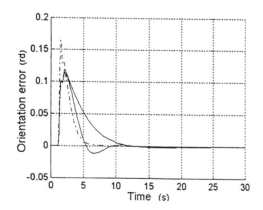

Fig. 12. Error in orientation
··· nominal plant ; — extremal plants

6. CONCLUSION

This article presents an additional application of the non-integer derivation in robotics (Oustaloup, 1995). After application to a tractor hitch system (Lanusse, et al., 1992) and then to a wire-guided robot (Sabatier, et al., 1995), the third generation CRONE control was applied here to an off-road mobile base with four steering and drive wheels. The strategy used makes it possible to take into account strong parametric variations on the one hand, and the non-linearities on the other hand, through the linearised tangent models calculated at non-zero speeds. The results obtained by simulation are remarkable, thus validating the strategy. Work on trajectory generation has already been published (Oustaloup, et al., 1994). Other work such as the generation of fractal terrains is on-going.

7. REFERENCES

Ackermann, J., W. Sienel and R. Steinhauser (1993). Robust automatic steering of a bus. In : Europ. Control Conf., Grenoble.

Canudas de Wit, C. and O.J. Sordalen (1992). Exponential stabilization of mobile robots with nonholomic constraints. In : IEEE/TAC.

Fossard, A.J. and D. Normand-Cyrot (1993). . In : Systèmes non linéaires (Ed. Masson).

Kanayama, Y., Y. Kimura, F. Miyazaki and T.. Noguchi (1991). A stable tracking control method for non-holonomic mobile robot. In : IEEE/RSJ Inter. Work. IROS, Osaka.

Lanusse, P., A. Oustaloup, C. Ceyral and M. Nouillant (june 1992). Optimal CRONE control of a tractor hitch system. In : IFAC, NOLCOS, Bordeaux.

Lanusse, P., A. Oustaloup and B. Mathieu (17-20 October 1993) Third generation CRONE control. In : IEEE/SMC'93 Conference, Le Touquet.

Oustaloup, A. (1991). In La commande CRONE, (Ed. HERMES).

Oustaloup, A. and H. Linarès (1994). The CRONE path planning. In : IEEE/SMC, Lille.

Oustaloup, A. (1995). In : La dérivation non entière, (Ed. HERMES).

Sabatier, J., A. Oustaloup, P. Lanusse and G. Robin (September 1995). Commande CRONE d'un chariot filoguidé. In : ICAR 95, Barcelone.

Wulfsohn, D. and S.K. Upadhyaya (1992). Determination of dynamic three-dimension soil-tyre contact profile. In : Journal of Terramech., Vol. 29, N°4/5, pp. 433.

PREDICTING THE USE OF A HYBRID ELECTRIC VEHICLE

C P Quigley, R J Ball, A M Vinsome.

Dr. R P Jones.

Warwick Manufacturing Group,
Advanced Technology Centre,
University of Warwick,
Coventry CV4 7AL, U.K.,
Tel: +44(0)1203 523794
Fax: +44(0)1203 523387
E-Mail: c.p.quigley@atcmail.warwick.ac.uk

Department of Engineering,
University of Warwick,
Coventry CV4 7AL, U.K.,
Tel: +44(0)1203 523108
Fax: +44(0)1203 418922
E-Mail: pj@eng.warwick.ac.uk

Abstract: This paper outlines the initial stages of a project to analyze the requirements and then design an intelligent controller for hybrid electric vehicles. Such a controller would be required to manage energy flow through the hybrid drive train and for optimum control would require a number of parameters normally available only upon journey completion. This paper presents work to attempt to predict these parameters at the start of the journey using intelligent classification techniques and a knowledge base of previous journey histories.

Keywords: Hybrid Vehicles, Prediction Methods, Automotive Control, Intelligent Control, Navigation Systems.

1. INTRODUCTION

Much of the research into new forms of vehicle propulsion is motivated by legislation intended to limit the polluting effects of vehicle exhaust emissions. Future legislation in both the USA and European Community will introduce progressively more stringent limits on vehicle exhaust emissions over the next 15 to 20 years. Ultimately, there will be a requirement for vehicle manufacturers to supply Zero Emission Vehicles (ZEVs) and Low Emission Vehicles (LEVs) for use as a form of private transport within large urban conurbations. At present electric vehicles are the only practical candidates as ZEVs, whilst hybrid electric vehicles currently form the most serious contenders as LEVs.

Hybrid electric vehicles have a propulsion system which includes a heat engine, and one or more electric motors and/or generators with an associated traction battery. The propulsion system in a hybrid electric vehicle can be assembled in a variety of configurations. One possible arrangement is a parallel hybrid vehicle consisting of a single electric motor (E) and heat engine (HE) mechanically coupled to a single drive shaft. A typical parallel hybrid vehicle configuration is described in figure 1. Power from the electric motor and/or heat engine is transmitted to the road wheels by drive shafts and gear mechanisms. Power for the electric motor is supplied by the traction battery (B). If the motor is driven either directly from the heat engine or during braking, it will generate current to charge the traction battery.

In its simplest form, the power from the two drives can be provided in one of three modes:-

1) Electric motor only.
2) Heat engine only.
3) Electric motor and heat engine combination.

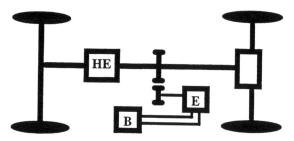

figure 1. Structure of a typical parallel hybrid vehicle.

Using the electric motor or heat engine exclusively (modes 1 and 2) present a manageable control problem for the driver of the vehicle, but their combined use (mode 3) makes it very difficult for the driver to control optimally. Previous work at the University of Warwick (Farrall and Jones, 1993; Farrall, 1993) has investigated the use of fuzzy decision making for the management of energy flow within a hybrid electric vehicle in this third mode. It was concluded that fuzzy control could provide benefits over a limited range of operation, but in order to obtain better performance over the complete range of operation, a method of adapting the fuzzy rules would be required.

To enable a hybrid electric power train controller to adapt to a wide variety of vehicle operation many parameters not normally used in vehicle control systems would be required , e.g. Journey duration, journey distance, time of departure, journey destination. Unfortunately most of these parameters are only known upon completion of a given journey. Therefore a means of intelligently estimating these parameters, based on the controller's past experience is needed.

A programme still very much in its infancy is the design of an intelligent controller. The proposed controller will allow journey parameters to be reliably estimated upon journey departure, and therefore allow for optimal operation with respect to exhaust emissions and fuel consumption.

The successful implementation of such a controller relies on the idea that many cars will have habitual usage characteristics for a high percentage of their journeys, and hence the ability to predict the occurrence of a journey and its associated characteristics will be quite high. A commuter journey is a particularly good example of a journey that exhibits habitual characteristics, and in the UK accounts for around 20% of all journeys taken by car (National Travel Survey, 1993). Other journey types (Business, Education, Shopping, Leisure) may also exhibit habitual characteristics, therefore the benefits to be gained from the use of an intelligent controller are high. The impact of such benefits will depend upon the geographical location the vehicle is used in.

For example in the UK, 16% of commuters in the London area drive their cars to work, whereas outside London almost 50% of commuters drive to work (National Travel Survey, 1993).

2. HYBRID POWER TRAIN CONTROL

As previously stated, the proposed controller will allow journey parameters to be reliably estimated upon journey departure. In order to do this the information available to the controller could take one of two forms:-

1st Generation Control
The essence of this type of control is that all information would only be available internally to the vehicle, from transducers belonging to the vehicle. A controller of this type, if implemented, would use signals derived from technology already present in modern day vehicles (e.g. electronic tachometer, engine management system).

Such information would include:-

a) Drivers Operational Inputs:-   Throttle
                                                 Brake etc.
b) Time of day/year.
c) Engine Management Data:-   Engine speed etc.
d) Road speed.

2nd Generation Control
This type of control would employ the use of 1st generation control information, but also would have the additional advantage of vehicle location information relayed into the vehicle from external sources. Such external sources could take the form of a GPS (Global Positioning System) navigation system, or a road transport telematic infrastructure of the future, perhaps employing the use of road side beacons. Much research is currently underway into the use of such systems for road transport. GPS systems have been suggested for use in vehicles for a variety of applications; for example rapid vehicle location in the event of a road accident (Voger and Harrer, 1994). Road side beacons have been a suggested tool not only for automatic debiting in road tolling schemes, but also for interactive route guidance systems (Bueno and Ongaro, 1991). A 2nd generation controller would explicitly know its location by receiving vehicle location information via such systems.

Information available:-
a)All 1st generation control information.
b)Vehicle location information i.e. latitude and longitude.

Control Decisions.
An intelligent hybrid electric powertrain controller would have to make a decision of control strategy in

two situations for the hybrid electric vehicle; <u>Where Am I Going?</u> and <u>Where Am I Now?</u>. The prediction of vehicle usage is concerned only with the <u>Where Am I Going?</u> decision. <u>Where Am I Now?</u> is a means by which an intelligent controller would continually reassess its original prediction.

<u>Where Am I Going?</u> is the initial estimation on the type of journey at the time of departure. The initial decision here is based on the Expectation of a journey type. Information available at the time of this decision differs between 1st and 2nd generation control.

<u>1st Generation Control</u>:- This can only use the present system time (time of day, day of week), the only information that is available at this instant. A journey is expected only if it frequently occurs at the same time of day, e.g. a morning commuting type journey.

<u>2nd Generation Control</u>:- This has the additional information of the vehicle's ground position at the time of departure. Expectation here is based on time of day, day of week and the vehicle's present global position.

If the controller decides a journey is expected it can make an estimation of the expected journey parameters, and an appropriate optimized control strategy can be referenced from the controller's memory. If a journey is not expected, the controller will choose the use of a general purpose control strategy, thus providing reasonably efficient operation only. Figure 2. shows the decision process.

Previous studies (Smeed and Jeffcoate, 1971; Herman and Lam, 1974) have examined commuter journeys in different geographical locations. They have found that they can describe the travel time variability on these routes mathematically with variables such as departure time and journey distance.

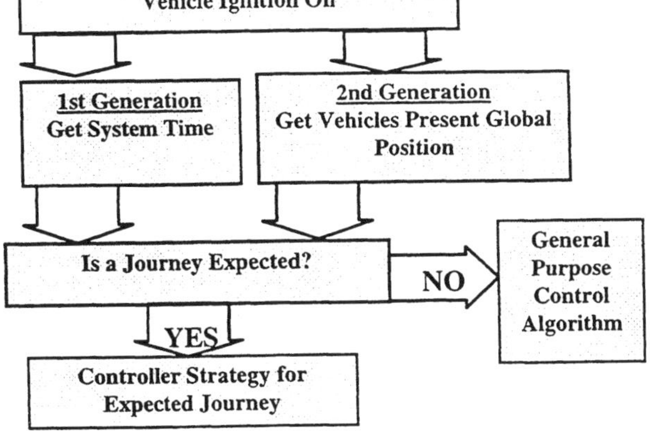

To the knowledge of the authors, the work described here involving the prediction of journey characteristics upon departure is unique.

3. VEHICLE EXPERIMENTS

Data is required throughout the project for investigation into methods of vehicle use prediction. This is achieved by the use of a data logger based around a GPS navigation system. Data recorded by the logger can be considered in terms of 1st or 2nd generation control data as follows:-

<u>1st Generation Control Parameters</u>
Time of departure,
Journey time elapsed,
Speed over ground, derived from latitude and longitude.

<u>Additional Parameters for 2nd Generation Control</u>
Explicit vehicle location (i.e. Latitude, Longitude)
Bearing relative to north, derived from latitude and longitude.

The GPS data logger is described in Figure 4. The GPS system requires signals from 3 satellites in order to obtain a 2-D position fix, and at least 4 satellites to obtain a 3-D position fix.

Careful consideration has gone into the selection of subjects. They are being chosen so that the main vehicle user represents a subset of the UK driving population, and their selection is based on age and sex statistics from UK driving licence registrations (National Travel Survey, 1993) and occupational statistics (Labour Force Survey Quarterly Bulletin, 1994). The subjects are being selected to give a spread of different occupations, different geographical location and different hours of work (shift, fixed hours, flexible time, part-time etc.). As a result, twenty subjects will have the GPS data logger installed in their car for a period of one month each.

Figure 2. Decision Process of Where Am I Going?

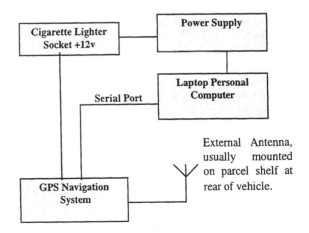

Figure 4. GPS Data Logger set-up

4. PRELIMINARY FINDINGS

At the time of writing, only the results from one vehicle were available. All of the journeys taken in this vehicle have been logged over a one month period; a total of 125 journeys. Figure 5 shows an example of the form of the raw data for latitude and longitude obtained on a single journey. Speed over ground and bearing are derived from this data.

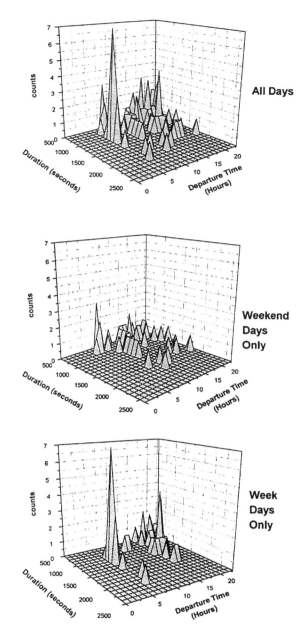

All Days

Weekend Days Only

Week Days Only

Figure 6. Variability of Journey Duration

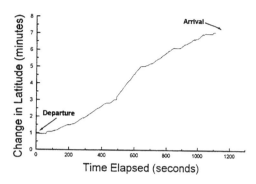

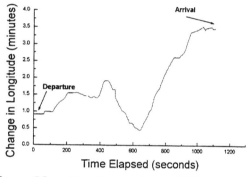

Figure. 5 Raw Data Obtained from GPS Data Logger.

A summary of initial data analysis is presented (i.e. journey duration, journey distance, and ground locations visited). The data is considered to develop rules to assist in journey prediction. 1st generation control data is considered first. This is followed by considering any additional advantages gained by using 2nd generation control.

1st Generation Control Data

As it is the case that many people work a five day week Monday to Friday, the data has been split into two distinct subsets for consideration; weekdays and weekend. Figure 6 shows the duration of our subjects journeys over the month period.

The weekend distribution shows a no obvious pattern, whereas the week day plot shows a number of journeys occurring between the hours 07.00 - 08.00, with a duration between 1000 - 1300 seconds. This actually corresponds with the subjects morning journey to work.

Figure 7 shows the distribution of the distance covered on the vehicle's journeys. Again the weekend distribution shows no obvious pattern. The week day plot shows a number of journeys occurring between the hours 07.00 - 08.00, with a distance around 14km. This correlates well with the data obtained for journey duration. The journeys occurring between 07.00 - 08.00 hours in fact account for only 13.6% of the journeys in one month.

From this data we can deduce the following rule on journey expectation:-

If it is a weekday, *and* the time is between 07.00-08.00 a.m.

then

there is a high expectation of a journey of 1000 to 1300 seconds duration, with a distance around 14km.

Although this information is useful, the distributions in figure 8 do not show how the locations are related.

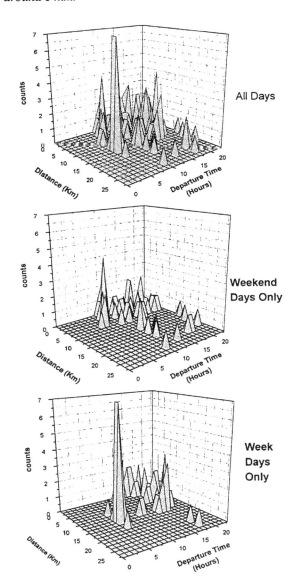

Figure 7. Variability of Journey Distance

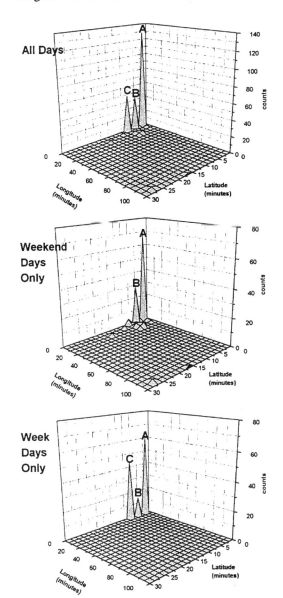

Figure. 8 Area Covered by Vehicle During Logging Period

2nd Generation Control Data

Plotting the cumulative distribution of vehicle global position (expressed in latitude and longitude) both at journey departure and arrival, gives an indication of the locations regularly visited. Figure 8 shows the distributions for all days, weekend days only, and week days only.

The weekend distribution shows a number of locations visited covering a large geographical area. Only two of these locations are regularly visited (Positions A and B). The week day plot shows only three visited locations (Positions A, B and C).

Nodal Analysis (Figure 9) gives a better view of how each location is inter-related. In the example shown, locations A, B and C account for about 62% of the journey destinations for this subject. Optimization from journey prediction could provide quite high benefits if the interconnecting journeys to these locations could be predicted. Each node has latitude and longitude as its properties (not shown on the diagram), each arc of the nodal diagram has the following properties associated with it:-

Length of journey.
Duration of journey.
Usual times of departure, e.g.

Day of week
Time of day

'Other Location' represents various other locations that have been visited during the logging period by the vehicle, but are not regularly visited, therefore appear unpredictable at this stage. 'Other Location', if unpredictable, would be a 'Don't Care' situation and would result in the use of a default control strategy in the final controller.

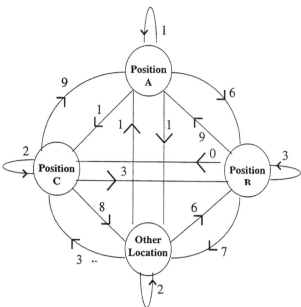

*The numbers at each arc represent the number of journey occurrences.

Figure 9 Nodal Analysis of Locations Regularly Visited for All Days in One Month Logging Period

Using all the data collected (i.e. both 1st and 2nd Generation control) the following rule can be deduced

If *it is a weekday,* **and** *the time is between 07.00-08.00 a.m.* **and** *I am at Position A*
then
> *there is a high expectation of a journey to Position C for a duration of 1000 to 1300 seconds over a distance around 14km.*

From the limited data collected, we cannot yet state reliably how much higher the expectation of a particular journey would be using 2nd generation control.

5. CONCLUSION

It is seen from the initial results from only the first subject, that it is possible to construct simple rules that could form the basis of an intelligent controller. This motivates the need for further research. Nodal analysis has shown that the first subject to take part in vehicle experiments predominantly visits just three locations, accounting for more than half of the journeys taken during the one month data logging

period. Nodal analysis of the locations visited by the vehicle, will be very useful if a 2nd generation controller is to be implemented (i.e. in terms of the number of nodes to be accommodated for), but ultimately we may not have the luxury of such a system. Initial results suggest that implementation of a 1st generation controller will be the most challenging, but this could be the least rewarding in terms of the number journeys that are predictable for a given vehicle. Further work will include a continuation of the vehicle logging programme and a more in depth examination of the data.

ACKNOWLEDGMENTS

The authors acknowledge the support provided by Rover Group for this work. The work was funded in part by the UK Engineering Physical Sciences Research Council (Grant No. GR/K35976).

REFERENCES

Bueno, S. and Ongaro, D. (1991), "Vehicle/Roadside Communication for Route Guidance", *Proceedings of the DRIVE Conference*, Brussels, **Vol.1**, pp4-6.

Farrall, S. D. and Jones, R. P. (1993), "Energy management in an automotive electric/heat engine hybrid powertrain using fuzzy decision making", *Proceedings of 1993 IEEE International Symposium on Intelligent Control*, pp463 - 468.

Farrall, S. D. (1993), "A Study in the Use of Fuzzy Logic in the Management of an Automotive Heat Engine/Electric Hybrid Vehicle Powertrain", *Thesis (Ph.D.)* - University of Warwick.

Herman R, Lam T (1974), "Trip Time Characteristics of Journeys To and From Work, *Proceedings of the Sixth International Symposium on Transportation and Traffic Theory*, pp57-85.

Labour Force Survey Quarterly Bulletin (Dec. 1994), The Government Statistical Service No.10, pp6.

National Travel Survey 1989/91, Transport Statistics Report, HMSO, London, September 1993.

Smeed, R. J. and Jeffcoate, G. O. (1971), "The Variability of Car Journey Times on a Particular Route", *Traffic Engineering and Control*, **Vol.13**, pp238-243.

Vogal D, Harrer S (1994), "DGPS - Emergency Location System for Vehicles", *The Journal of Navigation*, **Vol.47**, pp349-360.

AUTHOR INDEX